CHRISTIAN ENDEAVOR YEAR BOOK

CONTAINING
THE OFFICIAL REPORT OF THE
FIRST INTERNATIONAL

CHRISTIAN ENDEAVOR CONFRENCE

HELD IN

BUFFALO, NEW YORK

AUGUST 5 - 10, 1919

First Fruits Press
Wilmore, Kentucky
c2016

First Fruits Press
The Academic Open Press of Asbury Theological Seminary
204 N. Lexington Ave., Wilmore, KY 40390
859-858-2236
first.fruits@asburyseminary.edu
asbury.to/firstfruits

Christian Endeavor
YEAR BOOK

CONTAINING

THE OFFICIAL REPORT

OF THE

FIRST INTERNATIONAL

CHRISTIAN ENDEAVOR CONFERENCE

HELD IN

BUFFALO, NEW YORK,

AUGUST 5-10, 1919

UNITED SOCIETY OF CHRISTIAN ENDEAVOR
BOSTON, MASS.

CONTENTS.

CHAPTERS.

ILLUSTRATIONS.

OFFICERS

OF THE

United Society of Christian Endeavor

World's Christian Endeavor Building, 41 Mount Vernon Street, Boston, Mass.

PRESIDENT
REV. FRANCIS E. CLARK, D.D., LL.D.

ASSOCIATE PRESIDENT
DANIEL A. POLING, LL.D.

VICE-PRESIDENT
REV. HOWARD B. GROSE, D.D.

CLERK
MR. H. N. LATHROP

TREASURER AND PUBLICATION MANAGER
MR. A. J. SHARTLE

AUDITOR
MR. J. J. ARAKELYAN

GENERAL SECRETARY
MR. E. P. GATES

DEPARTMENTAL SECRETARY
MR. CARROLL M. WRIGHT

EDITORIAL SECRETARY
REV. ROBERT P. ANDERSON

EXTENSION SECRETARY
REV. IRA LANDRITH, D.D., LL.D.

SOUTHWESTERN FEDERATION SECRETARY
MR. W. ROY BREG

CHAIRMAN ALL-SOUTH EXTENSION COMMITTEE
LIEUT. DUNCAN B. CURRY

National Superintendents

ALUMNI
MR. STANLEY B. VANDERSALL

JUNIOR
MISS GRACE F. HOOPER

INTERMEDIATE
MR. PAUL C. BROWN

CITIZENSHIP
DANIEL A. POLING, LL.D.

RURAL AND FAMILY ENDEAVOR
MR. JOHN R. CLEMENTS

ARMY AND NAVY
REV. S. C. RAMSDEN

WORDS OF APPRECIATION.

THIS, the official report of the first International Christian Endeavor Conference, is as nearly complete in every detail as it was possible to make it. It represents not only the coöperative work of field-secretaries, conference-leaders, and the Buffalo newspapers, but chiefly that of Professor Amos R. Wells, managing editor, and R. P. Anderson, associate editor, of *The Christian Endeavor World.*

It is therefore with a keen sense of appreciation that the compiler heartily thanks the many friends for every available source of information and help necessary in the compilation of this report. They made it possible.

With the hope that this report will be a great blessing and inspiration to all who may possess it, and that success may attend every effort to promote Christian Endeavor, I submit this report.

A. J. SHARTLE,
Publication-Manager,
United Society of Christian Endeavor.

September 1, 1919.

THE BUFFALO CONFERENCE COMMITTEE.

Front row, left to right: George J. Barnett, Publicity; Agnes E. Baker, Entertainment and Banquet; Etta B. Fulton, Secretary and Billeting; Edythe M. Sawyer, Registration; Joseph C. Dale, Ushers; R. George Lord, Member Advisory Board.

Back row, left to right: Walter F. Ault, Excursion; Henry S. Gage, General Chairman; George E. Gowing, Choir Director and Music; J. Walter McElvein, Treasurer; Lloyd H. Roos, Churches, Halls, and Decorations; Frederick W. Fisher, Finance.

BUFFALO CONFERENCE COMMITTEE.

By A. J. Shartle.

THE International Christian Endeavor Conference at Buffalo was a wonderful blessing to Christian Endeavor. However, to make it the great success it achieved required a committee of exceptional ability; one imbued with a spirit of devotion to the cause, loyalty to its ultimate success, and self-sacrifice to a degree which can be measured only by practical experience.

All of this was embodied in the Buffalo Conference Committee, a committee of twelve consecrated Christian men and women who together, by their congenial fellowship and efficient work, made possible all of the many things Christian Endeavorers enjoyed at Buffalo.

It required many months of diligent work to provide for all of the essentials necessary to make the Buffalo Conference a success. Think of the host of preliminary meetings and loads of details, coupled with the anxious desire that all go well. Then add the sleepless nights when plans were reviewed and the net drawn closer in order to intensify the work during the last few days prior to the opening of the Conference. But the committee were on the job. *They did not fail*, and how proud we are of them!

When the Boston special arrived at the Union Station, Buffalo, in the early gray dawn of August 4, and I clasped the hands of that fine specimen of young American manhood, Mr. Henry Gage, general chairman of the Conference Committee, I knew that he and his committee were ready to meet every requirement. However, let us not forget that long before the Conference opened it required publicity that would attract a large constituency to the Queen City by the Lakes; and it was in this particular phase of work that the committee made a wise selection in the radiant, congenial, efficient Mr. Barnett. Of course we had good publicity, and consequently the large delegations that came required entertainment and food, and here again there was a wise selection in that dynamo of intensified energy and enthusiasm, Miss Agnes Baker, chairman of the Entertainment and Banqueting Committee, who was everywhere, any time, and all the time. Then there was Miss Fulton and Miss Sawyer, a splendid team who did fine work in billeting and registration respectively.

Mr. Dale as head usher kept everybody smiling and happy by giving each delegate and friends the best seats in the house, and we should not forget that Mr. Lord is the kind of man always helpful on the Advisory Committee. And shall we ever forget the excursions and the Canadian rally and banquet with all the attending features? We owe these pleasures to Mr. Alt, the efficient excursion-manager.

The singing, did you say? Was it not wonderful? And who but Mr. Gowing could have succeeded in producing so wonderful singing by that splendid chorus? It lifted those great audiences in Elmwood Hall to heights of delight that will long remain an inspiration to those in attendance. Gowing kept them going. McElvein was a courteous, business-like treasurer whom we learned to love, and Roos was indeed a marvel at securing the best places of meeting and decorations. Think of Roos and Gage decorating trolley-poles on Buffalo's Main Street at 2 A. M. the night before the morning after, and, finally, how could the Conference have been a success without money? So they put Mr. Fisher at the head of finance. So God bless you, Buffalo Conference Committee. You did magnificently on short notice. Our hats off to you.

Mr. Henry S. Gage
General Chairman Buffalo Conference Committee

CHAPTER I.

OFF AT BUFFALO.

WE arrived at the Queen City of the Lakes during the early dawn of a gray morn. Had we arrived hundreds of years earlier at the same place, a tribe of Kahkwas Indians would have received us in a basswood forest. There would have been huts and wigwams as places of entertainment. Fishing would have been in its glory, and Mr. High-Cost-of-Living a nonentity.

Instead, we were met by a smiling reception committee as the train stopped at a place where sometime there will be erected a magnificent union station. The basswood forest, the huts, and the wigwams of the long ago were no more; but rather there lay before us a municipality of the first magnitude, with the citizens waiting with outstretched arms to greet and entertain the Christian Endeavor hosts of 1919. What a change!

Amid piles of bags and baggage representing most of our earthly possessions we found our friends, our conveyance, and our places of entertainment. And just as in the long, long ago they had the Iroquois tribe in their midst, so we too met the Iroquois on its own ground, and they took us in—the hotel.

The experiences of the New England delegations were simply those of other delegations coming from every State in the Union and from the provinces of the Dominion of Canada. How splendid a spectacle, had it been possible to witness at a distance the Endeavor army converging on Buffalo! The modes of travel coupled with the outbursts of youthful religious enthusiasm were manifest all along the way, whether the deputations came from the north, the east, the south, or the west. And they came two thousand and seventy-eight strong.

They were the real cream of Christian Endeavor, and they came for business. They came with expectation, and went back filled with new methods and plans, new hope, renewed enthusiasm, and with a determination to give and do their best for God.

We did not expect so many delegates because this was to be only a *Conference*, the kind of meeting where the live wires of their respective organizations could make a through connection with the powerhouse of Christian Endeavor without grounding and consequently short-circuiting needful messages during this trying period of civic and religious reconstruction after a world war.

The First Baptist Church was the place of registration. It was here where during the hours preceding the opening of the Conference old friends met, and new friendships were formed. And it was not very long after the morning hours of August 5, when the fluttering of Christian Endeavor badges was seen in every section of Buffalo.

The United Society literature stand was here established, as well as in Elmwood Music Hall. It contained a feast of good things for the delegates who crowded the place between meetings in order to secure the latest and best helps on the work of Christian Endeavor. Mr. Walker and his efficient staff met every requirement of the delegates in a manner worthy of the highest commendation.

Part of the first day was spent in sightseeing. Buffalo contains many places of interest, and the groups of Christian Endeavorers strolling along the streets, or riding on sightseeing autos or trolley-cars were in evidence everywhere. Christian Endeavor took possession of Buffalo on the first day of its arrival, and held on to the last minute of its stay.

The meetings were wonderful. From the memorable opening session in Elmwood Music Hall when Dr. Clark gave the Conference his inspiring message, a message that was received with great enthusiasm, to the closing session which indeed was a real religious love-feast, every minute of the week of Christian Endeavor activities was one great blessing.

The conferences were of exceptional practical value; and, if one were to judge by the active participation on the part of delegates at the various conferences, challenge sessions, and denominational rallies, then surely these sessions alone exceeded in value a thousand-fold the financial investment made in sending delegates.

However, while these conference sessions proved their value from a practical point of view, we must remember that practice without inspiration often falls by the wayside. Consequently the Programme Committee wisely planned inspirational meetings for the evening sessions. Were they inspirational meetings? Ask the delegates and friends who were in attendance. Religious enthusiasm indeed was at a high pitch at every evening meeting. And the speakers? None better at any previous convention, conference, or rally.

Summarizing the Conference as a whole, it was a great success and far beyond our expectations. It was first of all an experiment. It was the first International Christian Endeavor Conference ever held. We did not know whether our constituency would approve it. It was too early and too close to the ending of the world war to attempt an International Convention with its more than ten thousand delegates. So we made it a Conference, and expected a small registration; and they surprised us with

2,078 delegates registered. This together with those not registered, and the friends in attendance made up one of Christian Endeavor's "best-yet" gatherings. Yes, the Buffalo Conference was a great success, and from a constructional, practical, and inspirational stand its value is immeasurable.

Finally, Buffalo as a convention city is second to none in the country. Locally, everything was done to make the visit of International Christian Endeavor a pleasure. Its beautiful streets, attractive buildings, and magnificent parks were a source of delight and admiration to all. Christian Endeavor has proved that it can not only hold a monster International Convention whenever it desires, but that it can also hold a conference on short notice, the like of which was never equalled. We are glad we went to Buffalo.

CHAPTER II.

A MEMORABLE OPENING SESSION.

ELMWOOD MUSIC HALL,

Tuesday Evening, August 5.

ELMWOOD MUSIC HALL, the leading meeting-place of the Conference, was most inspiringly decorated with great American flags hung thickly from the ceiling, with flags of many other countries grouped to emphasize the international charactcer of the gathering. The Intermediate white and green was also liberally used.

On the opening evening the Endeavorers came early to hear the fine organ recital by Mr. Irwin Binder. The organ, one of the largest and best in the country; was in the Music Hall of the Buffalo Exposition, and President McKinley was standing before it when he was shot. It was a gracious service on the part of Buffalo organists to take turns introducing the evening sessions of the Conference with recitals upon this noble instrument.

But the irrepressible spirits of the Endeavorers broke out, between the organ pieces, with their jolly cheers, the different States vying with one another in song and "yell." The red-coated delegation from Des Moines was especially vigorous in the front seats, but New York, Pennsylvania, Ohio, and many others were not behind in vim and originality.

As always, the entrance of Dr. and Mrs. Clark and the other national leaders was the signal for great applause, and the audience rose to receive them. A helpful praise service was conducted by Mr. George E. Gowing, the able Conference song leader, aided by the fine choir which Buffalo furnished, a choir exceptionally well trained.

Said Dr. Poling in opening the first session:

"More than any other Convention in the history of Christian Endeavor, this Conference is of vital and prophetic importance. The dark days of the Great War are behind us. The bright days of the opening peace are before us, filled with opportunity, filled with responsibility. This Conference is to show that in these days of peace Christian Endeavor will not be found wanting, as it was not found wanting in the days of war."

Dr. Poling introduced Dr. Clark, the presiding officer of the Convention, who was again greeted with great heartiness, and who received a spirited musical welcome from the choir.

The Trustees, Field Secretaries, and United Society Officers at Buffalo.

Dr. Floyd W. Tomkins, the beloved Protestant Episcopal trustee, conducted the opening devotional exercises, which were follow'd by a series of inspiring addresses of welcome.

Hon. George S. Buck, Mayor of Buffalo, said that he had thus greeted many conventions meeting in his city, but none had had so large numbers at the very opening session. His warm tribute to the influence of Dr. Clark was heartily applauded. He spoke of the anxiety of all public officials in these troubled days. "There is no jury fairer-minded than the American people," he insisted; and the path to safety lies in the right relations between the citizen and the State.

The counsellor of the Erie County Christian Endeavor Assembly, Rev. Frank Hyatt Smith, D.D., voiced the welcome of Buffalo's 124 churches and twenty denominations, "Martin Luther," he said, "had a tutor, and every morning the tutor lifted his hat to his pupil. 'Why do you do this?' asked Martin's father. 'I do not bow to the boy,' said the tutor, 'but to the possibilities of youth which he illustrates.' " In the same spirit, Dr. Smith said in his eloquent little address, he came to bow before the spirit of youth represented by the Christian Endevorers.

The thorough appreciation which we all felt regarding the preparation made by the Buffalo local committee showed itself in the splendid greeting which Chairman Henry S. Gage received as Dr. Poling introduced him,—"a large Gage but not a smooth bore," as Poling punningly said. Mr. Gage presented to Dr. Clark a Buffalo-made gavel in a neat speech, and then introduced his committee one at a time, each of them being vociferously applauded.

General Secretary Shaw was most enthusiastically received as he rose to read some of the special greetings that had come to him. The first was the following from the President of the United States:

"THE WHITE HOUSE.

"MY DEAR DR. CLARK:

"Your letter of July 23d gave me a great deal of pleasure and reassurance. I can assure you that your confidence and approval mean a vast deal to me. I dare not hope to be present at the Biennial Meeting to which you so generously invite me, but you may be sure that my heart will go out to them, and that I know what I am missing in missing the opportunity to address so great and influential a body on a matter so near my heart.

"Cordially and sincerely yours,

"WOODROW WILSON."

On the conclusion of this kindly message the audience rose spontaneously to show their gratitude and appreciation.

Secretary Shaw aroused tremendous enthusiasm as he read the spirited and pointed greetings that had come from China, Finland, Jugo-Slavia, South Africa, Mexico, the Maritime Provinces of Canada, Mayor Baker, of Portland, Ore., Bishop Fallows, Chinese Endeavorers of Chicago, and New South Wales. (See chapter on Messages of Greeeting.)

The chief feature of this glowing opening session was Dr. Clark's keynote address. The Endeavorers listened with eagerness to the words of their beloved leader as he spoke encouragingly of what Christian Endeavor has already accomplished, of its glorious services during the four years since we last met in an international convention, and of the great work for the coming years that is calling upon Christian Endeavor to do it. The great programme that Dr. Clark proposed to the societies was accepted and ratified by the representatives of the societies that crowded the hall. They will be true to these proposals during the coming years.

The address follows:

COME ON!

For New Victories from Old Ideals

It is four years since Christian Endeavor last met in international conference and Convention. We face a different world from the world of 1915. Scores of emperors, czars, kings, and princes have been deposed within the last eight and forty months. Autocracy has received a deathblow. A dozen new nations are knocking at the doors of civilization. The maps of Europe and Africa have been redrawn. The old geographies are as out of date as a last year's bird's nest or a Nineteenth Century almanac.

America has taken more advance steps than in a hundred years before the war. The shining seas on either side of her are no longer barriers but highways. The isolation of the olden days is forever ended. The air has become as familiar to us as to the birds. Birds of prey are those born of the war; doves of peace, carrier pigeons of good will, are ready for the better days to come. Prohibition has virtually become nation-wide. The Endeavor slogan of 1911 has been realized. Woman suffrage is no longer an iridescent dream. Woman has come into her own. Man has found his political and well as his social helpmeet. A League of Nations offers hope to a war-cursed world.

Truly a momentous quartet of years have been these last four.

The Church of God has not stood still. Mightier programs for the new era have been planned than any Christian dared propose four years ago. Dollars are asked to-day instead of pennies and dimes. Millions are demanded where thousands were once timidly pleaded for. The world calamity has opened the world's heart and pocketbook. Exclusive sectarianism is staggering to its fall. A league of churches as well as a League of Nations is a future outcome of the war. The Federal Council and the Interchurch World Movement are as inevitable as was an international army. A united Church under the one Captain of our salvation has been proved essential to victory. The vast denominational plans and "drives" summon the different divisions of the one army to go "over the top."

What about Christian Endeavor? Hundreds of thousands of our members in all lands heeded their country's call. Thousands of them lie unforgotten in foreign graves. Our societies and unions were for a time stripped of their leaders.

But Christian Endeavor was not vanquished. It is stronger than it was four years ago. New plans, with God's blessing, have resulted in new victories. New leaders have come to the front. Our veterans, the Alumni, are no longer on the shelf of retirement. They will henceforth occupy the front seats of honor. They will give us not only counsel and moral support, but the sinews of war as well, for new Christian Endeavor victories.

THE WHITE HOUSE
WASHINGTON

31 July, 1919.

My dear Dr. Clark:

Your letter of July 23rd gave me a great deal of pleasure and reassurance. I can assure you that your confidence and approval mean a vast deal to me.

I dare not hope to be present at the Biennial Meeting to which you so generously invite me, but you may be sure that my heart will go out to them and that I know what I am missing in missing the opportunity to address so great and influential a body on a matter so near my heart.

Cordially and sincerely yours,

Woodrow Wilson

Rev. Dr. Francis E. Clark,
President, United Society of Christian Endeavor,
Mount Vernon and Joy Sts., Boston, Mass.

What is true of American Endeavorers is true of most other countries. Great Britain has made notable advance in Christian Endeavor ideals and outreach. Continental Endeavorers have not been overwhelmed by war's horror. Australia and New Zealand Endeavorers, amid the greatest difficulties, have held their own. South America and Mexico have distinctly gone forward. China, Japan, and India have kept up their winning pace. Together they number their societies by thousands. Their members in these three nonchristian lands approach the 100,000 mark.

Here is a phenomenon to account for. The world has marvelously changed within four years past. Its politics, its statecraft, its governments, its methods of offense, and defense, its ideals, are not what they were, thank God! Some of them are different!

But the Church of God and its young auxiliary army have not changed their principles or purpose.

"Oh, where are kings and empires now,
Of old that went and came?
But, Lord, Thy Church is praying yet,
A thousand years the same.

"For not like kingdoms of this world,
Thy Holy Church, O God,
Though earthquake shocks are threatening her,
And tempests are abroad;

"Unshaken as eternal hills,
Immovable she stands;
A mountain that shall fill the earth,
A house not made with hands."

What is the reason for this stability, amidst the awful havoc of these four years? It is this: The foundation principles of the Church were laid by and in Him who is "the same yesterday and to-day and forever." Nations may welter in blood; politicians may scheme and fume and fail; the inventions of yesterday may be thrown on the scrap-heap to-day. Nevertheless, the foundation of God standeth sure. Our Lord's summary of the Ten Commandments—love to God and our fellow men—changes not.

To train young people to keep this commandment is the task of Christian Endeavor. Our pledge is this commandment in action, "I will strive to do whatever He would like to have me do."

Think for a moment what are the world's great present problems, problems born of the war. Are they not these?

1. To find efficient, unselfish leaders for the world's business and statecraft.

2. To find work for the unemployed.

3. To unite the nations in lasting fellowship and peace.

Our tasks, too, are to raise and train competent leaders, to find and assign to every one tasks worthy of a Christian, to unite all in a fellowship of service; these are the great tasks of the Church, and these are the tasks which Christian Endeavor is especially called to assume in these years of readjustment. Consider them one by one.

First, *our task is to raise up and train new leaders for the Church of God.* To this duty Christian Endeavor has addressed itself for eight and thirty years; to it we will devote ourselves more earnestly in the years to come. "Show us your men," said Phillips Brooks in his great sermon in Westminster Abbey. "Show us your men, strong men, and women, too, leaders of men and women," cries the Church to-day.

Christian Endeavor is not ashamed of her product. It has been and is a school for leaders. We will, with God's help, increase the output. Let us face the facts. Thousands of churches are dying for lack of leadership. Their membership is depleted to the lowest point. Not one new member is added in half a decade. The old gibe has some truth in it: "At Pentecost one sermon converted three thousand. Now it takes three thousand sermons to convert one." "Thou hast a name to live (a name written in the denominational year book), but art dead," is the verdict of the Spirit on other churches than Sardis. "Oh, for lay leaders, loyal and courageous!" is the unspoken cry of such churches.

But, brethren of the ministry and laity, this call is not met by any patent or any wholesale method. We may call out, as many do, for leaders; but echo answers, "Where?" How is this call to be answered? Not by giving easy tasks, or no task at all, to our members. It is not answered by simply opening the door of our Young People's Society so wide that worldliness, indifference, and carelessness can enter the active membership. The door to leadership is not unlocked by the key of a dance hall, a theater, or even a doughnut and a cup of coffee. The key to leadership is called "obligation, voluntarily assumed," in other words, a pledge to do "whatever He would like to have me do." He who adopts this covenant honestly, and lives up to it, and is endowed with ordinary tact and common sense, always becomes a leader.

I could call a long roll of the most eminent lay leaders of the day, and a list of thousands of ministers and missionaries, who will tell you that their first essay in Christian leadership was when they signed the Endeavor pledge. O pastors, show your young people how simple, yet how vital, how reasonable and how necessary, is such a pledge of loyalty and service. Keep them up to it, and you will have no lack of successors in the pulpit and of active supporters in the pews.

Thousands have already joined our legion of Life-Work Recruits. Hundreds of thousands more are waiting for your winning call to leadership.

Secondly, *unemployment* is the curse of the Church to-day, as it is of the industrial world. "Give us a job!" is the cry of idle millions. "Give us a task" is the unuttered, perhaps unconscious, demand of more millions in our churches. These millions are to be trained for service only by serving, for leadership only by leading. We hear many glittering generalities in these days about the eagerness of young people in the churches for a big job. "Give them a man's job, and men will be found to do it," it is often said.

But these same people who so insistently utter this cry never point out the man's job, but sometimes content themselves by declaring that Christian Endeavor does not offer one to its members. I beg leave to differ. Christian Endeavor is and always has been an employment bureau for the Church as well as a school for leaders. The big job always begins with the little job, faithfully, cheerfully, loyally, persistently performed.

The biggest job I can tackle is to strive throughout my whole life to do whatever my Master Jesus Christ would like to have me do. Is it not a task worthy of a moral hero to stand in the midst of a worldly generation for the things of the Spirit, to keep alive in the hearts of the young the love for the Bible and prayer and worship in God's house when a thousand siren voices call them away? Is it not an undertaking second to none in importance and difficulty to make religion the first thing in life, and everything else—school, sport, business—secondary? That is the high ideal and the tremendous task of Christian Endeavor. To do these things, this is a man's job, this is a woman's job.

Some people do not know a man's job when they see it. The young person who stands up for Jesus in the prayer-meeting is doing a man's job. It is of the very same nature as that of the boys who stood up for their country and her flag at Château-Thierry and Soissons, and at first it takes no less courage. To serve faithfully on the lookout committee or any other; to win a companion to Christ; to show him how to serve; to help a little Junior through the wicket gate; to guard a high school boy through the perils of adolescence—all these are men's jobs or women's jobs, for these two are but one. God hath joined them together in our society; let no man put them asunder by substituting classes of one sex for a united Young People's Society.

We rightly resent the implication that Christian Endeavor offers no worthy Christian tasks to young people. Yea, like Esaias the prophet, we would "be very bold and say" that it offers them the noblest, highest, hardest tasks.

To summarize—in the following ways our societies meet the peculiar demand of the Church in the new era for leadership, for employment, and also for fellowship.

Our pledge summons young people to Christian leadership.

Our prayer-meetings and our many committees train them for leadership by giving to every one his man's or woman's job.

And, thirdly, our union work—local, county-wide, state-wide, nation-wide, world-wide—gives the largest field and most abundant scope for fellowship, while maintaining our supreme loyalty to our own churches. "For Christ and comradeship over land and sea" was the noble motto of a British convention. It is one of Christian Endeavor's perpetual slogans.

By promoting this comradeship we again meet the challenge sometimes uttered by a friend, sometimes by a scoffer. "Let the Christian Endeavor movement give to the young people a worth-while task." Here it is: to unite millions of the young people of all denominations throughout the world in new and precious bonds of fraternity, common service, and unswerving loyalty to Christ and the Church. Can you think of a bigger job, of a more worth-while task?

That is what our societies are doing with ever greater success. They have already united young people of a hundred denominations in every land beneath the sun. The tie that binds is called a Christian Endeavor union, local, state, national, international. "Christ and comradeship" is the inner meaning of every Endeavor union from the Golden Gate to the Golden Horn, and from the Golden Horn around to the Golden Gate again.

A conservative estimate of our union meetings tells us that many millions of young people during the past four years, while the nations have been rent apart and the earth has been dyed red in fratricidal blood, have tasted the joys, the inspiration, and the uplift of Christian fellowship. These millions of voices have joined in the triumphant marching-song of Christian unity,

> "Like a mighty army
> Moves the Church of God.
> * * * * *
> We are not divided,
> All one body we,
> One in hope and doctrine,
> One in charity."

A league of the nations is on the way; a league of Christian young people is already formed. For this league of the young people of the churches also we draw our warrant from our pledge. This is one of the things *He* would like to have us do; for our Lord Jesus Christ has prayed in the most solemn hour of His life, "That they all may be one, as Thou, Father, art in Me and I in Thee, that they also may be one in us, that the world may believe that Thou hast sent Me." Jesus here implies that the faith of the world in Him depends on the unity of His followers. Is it not, then, a job worthy of any man—yea, is it not a tremendous Christian task—to promote this fellowship in a sectarian-rent, war-rent world?

If, then, leadership, employment for all, and fellowship among all, are the great ends to which we should address ourselves, what should be our goals and the road to them during the next two years? We have fought many bloodless campaigns in the past, campaigns inaugurated at such conventions as this; campaigns for increase, for efficiency, for millions. We thank God for them. All have succeeded. All the experiences of the past give us courage for the future, as we seek in the name of the Master and for His sake for more societies, larger societies, better societies.

We will challenge one another in this convention for these results, and thus provoke to love and good works. Two years from now we will celebrate the fortieth anniversary of Christian Endeavor. Thus far the Lord has led us on. That we may have during these two years simple, definite goals, by attaining which we may show our gratitude for the two-score years of divine favor, I suggest for your consideration these five:

1. *A fifty per cent. net gain in the membership of every society within two years.*

2. *A Christian Endeavor Society in every available church in every state, or, if not, we will know the reason why.*

3. *A definite goal in every state for denominational missionary gifts, especially in connection with the proposed denominational "drives" and interchurch movement.*

4. *A field secretary for at least part time in every state and province.*

5. *A sustaining alumni association in every union.*

Vigorous efforts to attain these goals will develop the leaders, will furnish the tasks, will promote the loyalty and the larger fellowship for which I plead.

If you agree with me in regard to these goals and the special needs of the day, you will also agree with me as to the supreme necessity of the renewed proclamation and promulgation of Christian Endeavor fundamentals in attaining them. We must show to our societies, our unions, and our churches the meaning, the wisdom, the necessity of the pledge, of the weekly meeting, of the roll call and consecration service, of systematic, comprehensive, persistent Christian activity. We will show not only how strong is Christian Endeavor in its underlying principles, but how flexible and adaptable to circumstances. We will affiliate with us, so far as possible, *all* the young people of the Church, while we will not lower the standards of *active* membership. We will keep our membership young, and lay the burdens of the society on young shoulders, since thus only can they be trained. We can do this as never before, since in the Alumni membership our older friends can still be related to us in sympathy and helpful activities.

To establish more firmly these essential, fundamental principles, I suggest that in every society and union we each year have one or more meetings or convention sessions devoted to the explanation, enforcement, and illustration of our pledge and our plans for religious activity and personal training for service and leadership.

A well-known missionary leader closed a ringing challenge to evangelize the world with two words, "Come on!" I, too, will adopt them. Come on for a great campaign of loyalty, loyalty to our churches and to these principles of Christian Endeavor: obligation, which is the heart of the pledge; unabashed, outspoken devotion, which is the heart of the prayer-meeting; systematic service, which is the heart of our committees; united service, which is the heart of our unions.

For pledged service, come on; for outspoken service, come on; for systematic service, come on; for united service, come on. Thus will leaders be raised up, great tasks be found and undertaken, our Lord's last prayer fulfilled. Then come on.

For our motto and our slogan can we have a better than the familiar words of our pledge, *"Whatever He would like to have me do"? "Whatever,"* a tremendous word! "Whatever," in speaking, doing, being, living, giving. "Whatever!" O Endeavorers: realize its breadth and length and depth and height. "Whatever," in church, in Sunday school, and in Christian Endeavor. "Whatever," in home, in community, in country. "Whatever," in business and politics.

Trusting in the Lord Jesus Christ for strength, I will strive to do *whatever* He would like to have me do. This is our great, our blessed, our joyous task. I summon you to undertake it. *Come on!*

The last half hour of this notable opening session was a whirlwind half hour, never to be forgotten. It was filled full to the brim by General Secretary Shaw. It was ostensibly only the giving of notices, but never before has the giving of notices been so gloriously magnified. Wit and wisdom, pathos and power, had equal parts in the treat he gave us. The delegates also, with frequent interruptions of handclaps and of State songs and impromptus, had frequent and varied part. Dr. Shaw said that soon he would be "sixty years young," and he proved that in spirit and vigor he is as young as ever. He announced his decision to resign from the office he has filled so long and with so great honor, an announcement which the Endeavorers emphatically disapproved and refused to accept; but he said he would remain a working Endeavorer to the last, and, when finally translated, would organize Christian Endeavor societies among the angels. Dr. Poling responded that the Endeavorers would compromise with him, and would accept his resignation, but only when he should begin his organizing work up above.

Thus ended one of the very best opening meetings our Society has ever held, a session afire with enthusiasm, warm with consecration, and inspired with lofty purpose.

CHAPTER III.

MEMORIAL MEETING.

First Baptist Church.

Wednesday Morning, August 6.

THE First Baptist Church was well filled on Wednesday morning for the short and simple but deeply meaningful meeting in memory of the Christian Endeavorers who died in the service during the Great War.

Dr. Clark conducted this memorial service most impressively It opened with the singing of that appropriate hymn, "The Son of God goes forth to war." Then we stood in silent prayer of thanksgiving for the noble lives of the dead, and of consecration to the high cause for which they died. These unworded prayers were earnestly voiced by Lieutenant Curry, chairman of the All-South Extension Committee, who has recently returned from the front.

Dr. Poling followed with an eloquent message. "The Great War," he said, "comes to me more and more as the Great Sacrifice, the sacrifice of treasure, of blood, and of youth." He spoke of that sacrifice, as he had twice seen it on the field in Europe. He pictured in living phrase the fidelity and unselfish heroism of the Endeavorers who fought in France or prepared to fight in the training-camps and were not able to get to the front. He pictured some of the stirring battle-scenes that he had witnessed, and he called upon the Endeavorers to devote their lives to the great aims for which our heroes died.

Amos R. Wells then read the following poem:

Discoverers.

By Amos R. Wells.

High glory his who walks where God alone
The mystic way has known,
  Who pierces first the mountain solitudes,
Treads first the echoing vaults of some vast caves,
Conquers the rage of undefeated waves,
  Or daringly intrudes
  Where immemorial arctic stillness broods
Above Death's timeless throne.
Praise, praise to him whose gallant mind
Knows how to find
  New roads of science, new domains of art,
New avenues of kingly thought,

Southern Delegation and United Society Officers.
Visiting Boston after the Buffalo Conference. Picture was taken in front of World's Christian Endeavor Building.

New mines whence happy myriads have brought
Balm to the senses, courage to the heart,
  Comforts to all mankind.
But glory, praise, and honor nobler far
To these whose guiding star
Rose in the east, and pointed them the way
To earth's most cruel fray,
Supreme of horrors, blackest pit of night,
War of the wrong and right.
These also with exploring feet have trod
Alone with God;
These also up to virgin heights have pressed,
  As ardent pioneers
  Have mastered fears,
And learned the wilderness by paths ungussed.
These, too, have reached the pole,
Have urged their dauntless soul
Through unimagined silences of snow
Where only night winds go,
Friendless and solitary and forgot
In that unhallowed spot.
No way that hero feet have trod alone
  Since the dim dawn of time,
  No venturing sublime,
But these young souls invincibly have known.
And they have found for us
Domains all-glorious,
Kingdoms of justice, empires of new good,
Sweet realms of brotherhood.
Yea, they have seen and caught
God's very central thought,
The truth of love supreme in sacrifice.
And they have paid the price,
The highest price wherewith the highest good is bought.
Massed in their swarming millions, each has walked
  In lonely places;
Each in his own high solitude has talked
  With angel faces;
Each has a separate conquest, and as each returns
His heart uniquely burns.
As each returns—but ah! we sing to-day
  Those who will not come back;
  We drape our flags with black,
And waft our mournful tribute far away.
And yet—are they not here?
For truth and freedom know not far or near,
The world is one
When glorious deeds are done,
And death itself is slain
By those that die a deathless end to gain.
Not in the sacred sod
Of battle-harried France alone with God
Are they asleep, but here, with God alive,
Their spirits gladly strive,
Uphold their proud beloved ones, proudly see
The world that they made free.
By every broken chain,
By every freedman, free man to remain,
By every darkened nation led to light,
By every baffled memory of wrong,
By every new-born permanence of right,

By every weakness learning to be strong,
Our fallen heroes rise,
Come from their graves with happy eyes,
And join the welcoming throng.
We clasp their comrade hands;
We catch from them the splendor of their mood:
Our spirit understands
What they have tested and have found it good;
And ours shall be with them henceforth to fight
For God and man, for liberty and right.

A fervent dedication service was then conducted by Dr Clark.

"This service," he said, "is not a service of eulogy, but of dedication; not merely a memorial of the dead, but for the consecration of the living. We can best honor our beloved Christian Endeavor comrades who gave their lives in the war for freedom and democracy by devoting our lives to the same great cause. The closing words of Lincoln's great address at Gettysburg were never more appropriate than to-day:

"'It is for us, the living, rather, to be dedicated here to the unfinished work that they have thus far so nobly carried on. It is rather for us to be here dedicated to the great task remaining before us: that from these honored dead we take increased devotion to that cause for which they here gave the last full measure of devotion; that we here highly resolve that these dead shall not have died in vain,—that the Nation shall, under God, have a new birth of Freedom, and that government of the people, by the people, and for the people, shall not perish from the earth.'

"Words without deeds are empty. Conflicts await us too. Swords may be beaten into ploughshares and spears into pruninghooks, but the fight against greed and selfishness and domineering, arrogant worldliness will never cease until He whose right it is to rule shall come.

"Shall we, comrades in Christian Endeavor, enlist in this war? If so, raise your right hands and say, 'THIS FOR ME,' as I lead you in this service of dedication:

"Our comrades warred not alone against mortal enemies, but against war and the fiendish cruelties of militarism. Shall we who remain enlist in this as yet unwon war? (*"This for me,"* said all standing with uplifted hands.)

"They fought that the submarine and the bombing-plane and poison gas and the noisesome trench and the spirit for which these things stand, might not embitter and dominate the lives of our children's children. Shall we fight so that these forces of the Evil One found in all lands shall not dominate the world? (*"This for me."*)

"They *died* that not might but right, as God gave them to see the right, should rule our land. Shall we *live* for the same great end?" (*"This for me."*)

"They dedicated their lives to the reign of the Prince of Peace. Shall we also dedicate our lives to the reign of righteousness under the Prince of Peace? (*"This for me."*)

On this high plane ended one of the most inspiring services that Christian Endeavor has ever held.

CHAPTER IV.

CORPORATION AND TRUSTEE MEETING.

First Baptist Church.

Wednesday Afternoon, August 6.

Corporation Meeting.

THE annual meeting of the United Society of Christian Endeavor Corporation was called to order promptly at 2 o'clock by President Francis E. Clark. There was a large attendance and much enthusiasm prevailed. Clerk Hiram N. Lathrop gave the order of business. After disposing of unfinished and general business pertaining to the corporation, reports were made by the officers. (See Chapter of Reports.) These reports proved the stability and steady growth of nation-wide and world-wide Christian Endeavor.

An interesting feature of this meeting was the presentation by Secretary Lehmann of the field-secretaries of the South, who have contributed so largely to Christian Endeavor success in this field.

Mr. Roy Breg, general secretary of the newly-formed Southwestern Christian Endeavor Federation, which comprises Texas, Arkansas, Colorado, Utah, Arizona, New Mexico, and southern Wyoming, briefly described plans in mind for the work.

Mr. Stanley B. Vandersall, appointed to be national Alumni superintendent, was presented to the audience and spoke words of hope for the work upon which he soon will enter. In the same way Mr. E. P. Gates of Illinois, who will take up work as general secretary of the United Society, spoke a few words in introducing Mr. F. D. G. Walker, who becomes field-secretary for Illinois. Mr. R. A. Walker of Indiana, who now has charge of the United Society's Chicago office, was also happily introduced.

Following officers were elected:

President, Rev. Francis E. Clark, D.D., LL.D.; Associate President, Rev. Daniel A. Poling, LL.D.; Vice-President, Rev. Howard B. Grose, D.D.; Clerk, Mr. H. N. Lathrop; Treasurer and Publication-Manager, Mr. A. J. Shartle; Auditor, Mr. J. J. Arakelyan.

Annual Meeting of the Board of Trustees.

HOTEL LENOX,
BUFFALO, N. Y., AUGUST 6, 1919.

The Annual Meeting of the Board of Trustees of the United Society was held in the Committee Room of Hotel Lenox, Wednesday, August 6, 1919, at 3:30 P. M.,, President Francis E. Clark in the chair. Prayer was offered by Dr. J. T. McCrory.

The minutes of the last meeting were read and approved.

The following officers, secretaries, and superintendents were unanimously elected by ballot:

Officers.

General Secretary, William Shaw, LL.D.; Publication Manager, Mr. A. J. Shartle; Field-Secretary, Mr. E. P. Gates; Editorial Secretary, Rev. Robert P. Anderson; Extension Secretary, Rev. Ira Landrith, D.D., LL.D.; Southwestern Secretary, Mr. W. Roy Breg.

National Superintendents.

Alumni, Mr. Stanley B. Vandersall; Junior, Miss Grace F. Hooper; Intermediate, Mr. Paul C. Brown; Citizenship, Daniel A. Poling, LL.D.; Rural and Family Endeavor, Mr. John R. Clements; Army and Navy, Rev. S. C. Ramsden.

The following were unanimously elected to serve as an

Executive Committee.

Rev. Francis E. Clark, D.D., LL.D., Rev. Daniel A. Poling, LL.D., Mr. A. J. Shartle, William Shaw, LL.D., Amos R. Wells, Litt.D., LL.D., Rev. Robert P. Anderson, Mr. Stanley B. Vandersall, Henry B. McClusky, M.D., Mr. J. J. Arakelyan, Rev. James L. Hill, D.D., Rev. Howard B. Grose, D.D., Mr. Charles H. Jones, Rev. F. G. Coffin, D.D., Rev. F. M. Sheldon, D.D., Mr. H. N. Lathrop.

After discussion, it was voted that we plan for a great Convention in 1921, to celebrate the fortieth year of Christian Endeavor, and that plans and preparations be referred to the Executive Committee, with power.

VOTED, That the question of a World's Conference in Europe in 1920 be referred to the executive officers, with power.

After a presentation of the budgets from China, India, Japan and Latin America, in connection with the Interchurch World Movement, the following resolution was presented by Dr. R. W. Miller, and unanimously adopted:

WHEREAS, The United Society of Christian Endeavor has received appeals from foreign mission fields to enlarge the work of Christian Endeavor, and the needs are urgent,

Part of that Splendid Conference Choir.

AND WHEREAS, Christian Endeavor is interdenominational, and its work builds up the local church, develops evangelistic workers, and Christian Endeavor extension can be carried forward with least expense to the churches,

THEREFORE BE IT RESOLVED, That a committee be appointed to confer with the several denominational mission boards to consider plans of coöperation in the work of Christian Endeavor in foreign fields.

The following motion was unanimously adopted:

VOTED, That a survey commitee of seven be appointed to cover the Christian Endeavor field at home and abroad; that this committee report a programme of service and a financial programme to the Executive Committee not later than December, 1919; and that the executive officers confer with the leaders of the Interchurch World Movement, presenting the results of the survey, and establishing the relation of Christian Endeavor to the movement.

The following committee was named to consult with the mission boards, and to make a survey of the field at home and abroad:

Rev. Rufus W. Miller, D.D., Rev. A. W. Halsey, D.D., Rev. James L. Barton, D.D., Rev. Gilbert Glass, D.D., Rev. G. W. Haddaway, D.D., Rev. Howard B. Grose, D.D., Rev. S. Guy Inman.

VOTED, That the goals suggested by Dr. Clark in his keynote address be heartily approved, and that the question of percentage of increase in membership be referred to the field-secretaries for consideration and report at the trustees' meeting Friday afternoon.

Meeting adjourned to Friday at 2 P. M.

WILLIAM SHAW,
Secretary.

FRIDAY, AUGUST 8, 1919.

The adjourned meeting of the Board of Trustees was held at 2 P. M., August 8, 1919, in the Committee Room at Hotel Lenox, President Francis E. Clark in the chair. Rev. Dwight L. Goddard of New York, a newly-elected member of the Board, led in prayer.

This was followed by a helpful discussion of Prison, Floating, Army and Navy Endeavor, Rural Work, and societies in schools and institutions.

VOTED, That the general secretary be requested to emphasize and stress during the coming year the work referred to above, working through our State and city unions.

VOTED, That the Board of Trustees request the Railroad Administration to grant special clergy rates to our full-time field workers, and that the general secretary present the appeal.

VOTED, That on recommendation of the field-secretaries, the Monthly Service Themes and Programme for Christian Endeavor Week be adopted. (See chapter on reports for copy.)

VOTED, That the Standards and Goals for 1919-1921, as suggested by Dr. Clark in his keynote address, be approved.

VOTED, That the departments of our work be unified and the campaigns be directed by the United Society.

VOTED, That the whole question of departments and superintendents be referred to the executive officers, in consultation with the field-secretaries, and report if possible at the next meeting of the Executive Committee: and that special consideration be given to social, recreational, and educational work.

VOTED, That the Alumni superintendent be requested to present at the next meeting of the Executive Committee goals and challenges for the Alumni, and that a month in the spring be set aside for a nation-wide Alumni campaign.

VOTED, That Mr. Barclay of the Boy Scout Movement, be invited to come to Buffalo as the guest of the Conference.

VOTED, That the following recommendation be adopted:

We recommend the organization of the Standards Program Commission of the United Society of Christian Endeavor, which shall consist of the executive officers of the United Society, together with the Young People's secretaries of the various denominations who are trustees of the United Society. This Commission shall take original and final action in all matters involving the relation of the societies to their denominations, such as the adoption of standards and programs of work. The Commission shall hold at least one meeting a year, for consideration and adoption of the Young People's program of the coming Church year.

Resolved, That it shall be the policy of the United Society of Christian Endeavor to maintain and invite the closest relations with the state Christian Endeavor unions. To this end the executive officers of the United Society will, whenever possible, consult with the state officers before arranging extended tours and campaigns in their respective states.

At this point representatives from cities desiring to entertain the 1921 Convention were introduced. The invitations were presented in alphabetical order. *Des Moines* was represented by Mr. Graves and Miss Alma Garber, who presented the claims of their city in a forceful and interesting way. The claims of *New York City* were presented by Rev. G. S. Bolsterle and Mr. E. F. Garwood. *St. Louis* was next introduced, and a strong appeal for the Convention was made by Mr. H. F. Hahn, which was supplemented by the assistant secretary of the Convention Bureau.

While the Committee were conferring with the New York delegation, with reference to its financial guaranty for the next Convention, William Shaw read his resignation as general secretary, a copy of which is appended to this report. He supplemented it with an earnest appeal that it be accepted. His associates and several trustees expressed their sincere regret and appreciation of his work, but out of deference to his wishes it was

VOTED UNANIMOUSLY, That Dr. Shaw's resignation be accepted. This was passed by a rising vote in token of love and respect.

VOTED UNANIMOUSLY, That a committee be appointed to draw up resolutions expressing the deep feeling of the Trustees in connection with the resignation of Dr. Shaw. The resolutions are appended to this report.

VOTED, That the 1921 Convention be held in New York City, on condition that New York provide the necessary financial guarantee by October 1, 1919.

Meeting adjourned till 10:30 P. M.

WILLIAM SHAW,
Secretary.

FRIDAY EVENING, AUGUST 8, 1919.

The adjourned meeting of the Board of Trustees was held this evening at 10:45, President Francis E. Clark in the chair.

Dr. Poling made a statement concerning the 1921 Convention. It was then

VOTED, That the result of the secret ballot taken with reference to a second choice, if New York fails to secure the necessary guarantee of financial support, such as the other cities gave, be declared. It was also definitely stated that this vote should not in any way affect the choice of the 1923 Convention city. Dr. Clark reported the result of the secret ballot, as follows: St. Louis, 25; Des Moines, 35.

VOTED, That the next meeting of the Executive Committee be held the first Wednesday in October.

VOTED, That General Secretary Shaw be made a life-member of the Board of Trustees.

VOTED, That in these days, when woman is coming into her own, Mrs. Francis E. Clark and Mrs. William Shaw be elected life-trustees.

VOTED, That the following summary of our standards and goals for 1919-1921 be heartily approved: (See "Loyalty Campaign.")

VOTED, That Mr. Paul C. Brown be requested to put in writing his suggestions with reference to social and recreational work, for reference to the Executive Committee.

Meeting adjourned, subject to the call of the president.

WILLIAM SHAW,
Secretary.

SUNDAY, AUGUST 10, 1919.

A special meeting of the Board of Trustees was held in the Elmwood Music Hall at 5:30 P. M., President Clark in the chair.

Mr. Barclay of the Boy Scout Movement, presented a plan for an alliance between the Boy Scouts and Christian Endeavor, with a representative of Christian Endeavor on the Board of National Commissioners. It was

VOTED, That we approve of the proposition, and refer it to the Executive Committee, with power.

VOTED, That we request Mr. Barclay to present our petition for recognition of service in Christian Endeavor in the Standards for Boys, now being prepared by a special commission.

Meeting adjourned.

WILLIAM SHAW,
Secretary.

Meeting in Elmwood Music Hall. Address by Secretary of War Baker.

CHAPTER V.

A WAR AND PEACE MEETING.

Elmwood Music Hall,

Wednesday Evening, August 6.

LONG before the hour of opening for the evening meeting in Elmwood Music Hall on Wednesday night rivers of young people wearing the bright Christian Endeavor badge made their way to the great hall where George E. Gowing, the choirmaster, was leading choir and audience in song. Never was better singing at any meeting. Let three thousand young people loose on "Brighten the corner," and there is a brightness that may be felt. Just to be in such a great company of youth is in itself an inspiration. The man that can attend a convention like this and despair of the future of the church is surely more than blind.

Dr. S. H. Woodrow of St. Louis, led the devotional exercises.

The Buffalo Endeavorers had prepared a patriotic pageant which was introduced by the choir's singing, "The Battle-Hymn of the Republic," while two representatives each of the Y. M. C. A., Salvation Army, Red Cross, marines, sailors, aviators, and soldiers filed on to the platform. The choir then burst into "The Star-Spangled Banner," the soldiers and others saluting, while Columbia herself, a flag-draped young woman, took her place in the centre of the group. As the others filed away, Columbia moved gracefully back and disappeared in the background. It was simple, but it was tremendously effective; for without a word being spoken the pageant told the dreadful story of the tragedy of war.

Dr. Clark presided over the meeting and introduced representatives of Christian Endeavor soldiers and other workers in overseas service.

Lieutenant Duncan B. Curry, chairman of the All-South Christian Endeavor Extension Committee, was the first to speak, interrupted, happily, by song and cheers from groups of Dixie delegates. Mr. Curry told of the difficulty of making soldiers of raw recruits, and claimed that we need to apply the principles of army training to Christian Endeavor and the church. If making soldiers is hard, making Christians is still harder; but one way to make them is to put upon their consciences the obligations of the Christian Endeavor pledge. The speaker took us to France and vividly described the struggle at Chateau-Thierry and Belleu Wood, when our men went up a steep ascent in the face of streams of bullets and captured the positions they went out to take. This is the spirit Christian Endeavor must have, the spirit of tackling hard tasks as those boys did who nobly faced grim death.

Rev. Frank Day, D.D., of Canada, a trustee of the United Society, was the next speaker. He began with a story of a man who tried to explain the phrase, "The sun never sets upon the British Empire," by declaring that God could not trust the marauding British in the dark. If that man lived to-day he would revise his opinion, for Britishers and Yankees have fought the battles of civilization and liberty together. He took us with him in thought to the famous and bloody Vimy Ridge. We saw the British Tommy, underfed, perhaps, a trifle, but with the same look of freedom and sense of responsibility in his eyes as we see in the eyes of our own men. The speaker forcefully laid upon all hearts the need of responsibility on the part of Endeavorers. For the Endeavorer is not like the soldier in Europe who, when asked what he was doing, said that he was looking for Admiral B.'s goat. "But Admiral B. has been dead for thirty years." "Yes, and so has his goat, but I have not received any other orders." Endeavorers do not cling to past orders; they are alive to the orders of the day, absolutely up to date.

A well-loved Endeavorer, Walter D. Howell, cited for bravery while serving in the Y. M. C. A., brought a thrilling message on the A. E. F. slogan in France, "Let's go!" It was the spirit of this overseas slogan that carried our boys to victory and scared the Germans to death. The audience listened breathlessly as he told of his experience with th First Division, first to arrive and fight, and last to leave. Vividly he brought before us a village drenched with gas, where the Y. M. C. A. canteen ran only from dark to daylight. The speaker told of gas-stricken men staggering to an aid station for help for their comrades in the trenches. The officer ordered ten men to bring in the wounded, but fifteen rose to go. The sergeants took the mask from his face for an instant and said, "Let's go!" They went. But they could not see with their masks on. One by one they took off their masks and the unmasked men led the others, holding hands, out and back. Every man of the ten was gassed. This is the spirit of the army, and it is the spirit of Christian Endeavor. Comrades of the Christian Endeavor army, "LET'S GO!"

Clarence C. Hamilton, field-representative of *The Christian Endeavor World,* who has been in Y. M. C. A. work in France, brought a stirring message lifted out of personal experience. He gave some striking facts about the work done by the Y. M. C. A., for the "Y" bought coal to warm the people suffering from cold,

Clarence C. Hamilton,
Field-Manager of the *Christian Endeavor World*.

organized factories to manufacture writing-paper, and did multitudes of other tasks for our own men and the people of devastated lands. One began to catch visions of a tremendous, unselfish service done by the "Y" for our boys "over there."

One of our aviators, Lieutenant James Wray, who was greeted by his Iowa friends with the yell, "*Hoo*ray, *hoo*ray, *who's* Ray? *James* Wray," made happy allusions to his experiences. He had determined, he said, that he never would take another flight, "but oh, my," he added, "I'm sure up in the air now!" He claimed that if he was a little stoop-shouldered (which he wasn't) it was from ducking under the clouds. Mr. Wray arrived at the front on the first day of the Argonne drive, in which he took part and of which he told in that witty vein that seems to go along with the invincible optimism of youth. Take his answer to the criticism which some fellows make that there wasn't an American plane at the front, "Tell that fellow that in the next war he must put a sky-light in his dugout."

Arthur E. Whitney of Des Moines, Ia., who is going to be the field-secretary of the Ohio union, and who arrived in this country four days before the convention, related some vivid incidents from "over there."

Mr. Carlton Sherwood, a New York boy, who was greeted with cheers by his townmates, told of many thousands of Endeavorers who are coming back to this country filled with a new spirit of enthusiasm, service, and sacrifice.

All these talks were "real stuff," the material out of which life is made. This quality made them a delightful feature of a delightful gathering.

There was competition among the State delegations to do honor with cheers for Daniel A. Poling, whose topic was "Christian Endeavor at the Front in War and Peace." It was an eloquent if short address, which did heart homage to the men who went away with the colors and who come again no more. Dr. Poling yielded most of his time to Hon. Newton D. Baker, Secretary of War, the man who did tremendous things to keep our boys fit and trim to fight the foes of liberty.

Secretary of War Baker, a man who, Dr. Clark said, loves peace better than war, but who, when war came, was able to equip millions of men to fight for civilization, received a genuine Christian Endeavor ovation, first when he entered the hall, and second when he arose to speak. A large delegation from the District of Columbia and other delegations, especially that from his own Ohio, greeted him with yells and song.

Secretary Baker is a quiet talker, but an eloquent speaker nevertheless, with perfect command of language, and he is a chooser of fit and beautiful words. His address dealt with the work of America in the war, but the speaker paid warm-hearted homage to the spirit and work and sacrifice of our Allies, for no tribute, he said, that can be paid to the Allies can possibly exceed the truth. America came late. For half a century Germany had been preparing to show that might is right, and when she fell upon the world it took the united strength of the free nations merely to hold the Germans in check. America came late, but she was doing great things toward the end. On the day of the armistice America had more soldiers in France than Great Britain had at that time. We were sending across danger-infested seas 300,000 men a month. And America's effort was on a rising scale.

Then, when the fighting ceased, we reversed the process, and in a few months brought back hundreds of thousands—nearly all—the men who went forth to sacrifice themselves for liberty.

Secretary Baker paid tribute to General Pershing, whose military imagination, he said, is of titanic character.

There are other features of our work, however, besides military features. When Secretary Baker took up his work he knew that we had no right to take young men from fathers and mothers, sweethearts and wives, homes and churches, and place them where they might contract other scars than those that might be received in honorable warfare. An efficient army must be a moral army. And Christian Endeavor was one of the first religious organizations which came out to help to keep our men clean and strong. The consequence is that the American soldier in France fought like a soldier and lived like a gentleman.

The speaker told of the service of the homes and the churches of America, and of the service of doctors, surgeons, and nurses for the boys, a service rendered often at immense sacrifice. He pointed out that there are five great enemies of mankind, disease, crime, insanity, drunkenness, and poverty. Secretary Baker showed that our army was the healthiest that ever fought,—healthier than the same number of men in civil life. Crime was practically eliminated. Instead of there being 200,000 men in military prisons, as would have been the case with the same number of men in civil life, there were in prisons only 2,000 men, most of them for petty offenses and for short terms.

There was little drunkenness. There were practically no cases of delirium tremens. And, in spite of the unusual strain upon the men, insanity was only two-thirds of the normal expectancy and experience in civil life. Our men went abroad chivalrous knight-errants, moral agents, seeking to rescue the liberties of the world.

When peace came, our leaders began to wonder what to do to help the soldiers to while away time before they could be brought home. The leaders thought that sport and athletics would be the things to provide; but the army answered the question for itself. The thing the army called for was *education,* and schools of all kinds were established in every division.

SECRETARY OF WAR BAKER.

But what has been done for the soldier can and must be done for the next generation of young men and women. We have proved that we can do it, and now we must never lose our grip upon the gains that the war has brought to us.

Secretary Baker graphically depicted the cost of the war in money—not to speak of other waste—when he said that if all the land, and the improvements upon it—*all* the wealth in the United States, were to be sold at a fair value, it would bring less money by ten billion dollars than the amount the war has cost the nations of the earth. That gigantic sum is figured by our Government at $186,000,000,000. The war has impoverished the world. The world has been widowed and orphaned, and now it is being starved. Armies have swept over once prosperous nations. Disease, epidemics, and hunger harry countless millions of people, and the foundations of order have been shaken. Men and women who are suffering as parts of Europe are suffering, have neither the faith nor the strength to build again a free civilization.

The eyes of those nations bend across the sea. They look to America where a free republic has been established, an example which these people may imitate. We are called upon to be leaders in a movement to make impossible a repetition of the catastrophe which has come upon mankind.

Mr. Baker referred to the Peace Treaty and claimed that even if it is imperfect, it is the only one that has been proposed to prevent a recurrence of war.

What is the alternative? The old way and the old end. If another war comes, the ingeniousness of men will make it vastly greater in its destructiveness than this war has been. If we adopt the League of Nations, we have at least a basis of progress, and there will grow up a firm and consolidated spirit of unity, based on the brotherhood of man and the Fatherhood of God.

A delegate from the floor proposed that the gathering send to President Wilson and the Senate a resolution in favor of the League of Nations, and this resolution was passed without a dissenting voice.

CHAPTER VI.

THE CHALLENGE CONFERENCE.

FIRST BAPTIST CHURCH,

Thursday Morning, August 7.

FRESH and bright and eager for more the Endeavorers streamed to the Quiet-Hour service on Thursday morning in the First Baptist Church. This meeting was led by Rev. G. W. Haddaway, D.D., president of the Methodist Protestant Christian Endeavor union, who spoke of the manifestation of the glory of Christ in social life at the wedding feast in Cana; of His sympathy for the host of the occasion, a sympathy that enters into even the little perplexities and embarrassments of life; and of the greatness and character of His power, a power that was used to help and serve and change men and women into the image of Himself.

An interesting part of the Conference was a statement of aims by the States for their work in the next two years. This was one of the snappiest meetings of the whole Convention. Endeavorers have learned not only how to speak enthusiastically, but also how to put their thought briefly in burning words. The aims, as we have put them into cold type, fail altogether to express the fire and the enthusiasm with which they were spoken.

Massachusetts promised to live up to the five goals presented by Dr. Clark.

Illinois's aim is to organize every one of the 102 counties in the State and to run a poster campaign to advertise Christian Endeavor everywhere in the State.

Wisconsin proposes to live up to the words of its seal, "Wisconsin for Christ," as well as to advertise Christian Endeavor in out-of-the-way places.

Pennsylvania has been four months without a field-secretary, as Mr. Macrory, the former secretary, has been called to a pastorate in Ohio. But Pennsylvania has gone to Ohio to get a new secretary, Mr. Haines A. Reichel, who will strive to perfect the organization in all counties in the State.

Colorado promises to double its membership, and in every direction increase its efficiency by one hundred per cent.

Florida has three aims: to survey the State and establish the possibilities for organizing new societies; to organize societies in these places; and to persuade every society in the State to use the Efficiency Chart.

Iowa will coöperate in work to reach the five goals presented by Dr. Clark; will seek personal touch with every society; and will, it hopes, entertain the next International Convention in Des Moines.

Mississippi also accepts the five goals, and will emphasize deeper consecration to Christ.

Nebraska expressed its aim in its motto, "Pushers pushing," and will push every phase of Christian Endeavor work, including the Alumni Association.

Minnesota's aim is to improve Junior, Intermediate, and Young People's work, as well as the efficiency of district unions.

Arkansas promised to endeavor to plant the Christian Endeavor banner in every part of the State, and especially to labor for increase and efficiency in Junior and Intermediate work.

Kentucky's purpose is to increase the membership of Junior, Intermediate, and Young People's societies.

North Carolina purposes to place information about this International Conference in the hands of every society by means of a flying squadron which will make an intensive campaign for this object. The State proposes to have extra field-work this year, and to make every society an efficient society.

Texas promised to try to do things that will compare with the magnificent size of the State. She proposes also to make the State the Big Sister of the new Southwestern Federation of Christian Endeavor, and to render financial and unselfish service to the other States.

Maine's purpose is to have a bigger budget than ever this year, to inaugurate a correspondence course in "Expert Endeavor," and to continue to make its summer school one of the best.

The conference that followed, a Challenge conference, is a new feature in our conventions. The idea was to have the State and city unions challenge one another to contests along Christian Endeavor lines, increase, efficiency, missionary interest, and so on. Dr. Clark was the leader.

Although the unions were handicapped by the fact that the idea is a new one and the Endeavorers had come to the Convention without having prepared their challenges, the idea at once caught on and fired the imagination of the young people. The beginning made in the conference promises that as the months go on an increasing number of States and local unions will send challenges to one another. As it was, the friendly rivalry manifested, and the loyalty of each State to its leaders and its work, were a revelation of the immense possibilities in this new feature of our work.

Challenges came thick and fast, and as they were flung out were ardently accepted by the States challenged.

Massachusetts challenged Iowa to a 50-per-cent gain in 100-per-cent Junior, Intermediate, and Young People's societies according to the standard set in the Efficiency Chart, and to organize Alumni Associations in local unions.

Arkansas challenged North Carolina to the observance of the monthly service programme of the United Society, and Florida to an increase in the number of Intermediates.

Illinois challenged Pennsylvania on new Intermediate societies; challenged Ohio on new Junior societies; challenged Minnesota on mission study; Iowa on new societies; and Texas on the largest number of subscriptions to *The Christian Endeavor World*.

Rhode Island challenged New Hampshire on the number of new societies organized, and Vermont on the largest increase in the number of Efficiency Charts used in societies.

Florida challenged Maine, Nebraska, Oregon, and California on the percentage of increase in the number of new societies organized, on the percentage of societies using the Efficiency Chart, and on the largest number of societies above 300 per cent according to the standard outlined on the chart. North Carolina and Nebraska at once accepted Florida's challenge.

North Carolina challenged Arkansas on the largest number of societies organized and on securing Comrades of the Quiet Hour; challenged Virginia on the Tenth Legion, Alabama on securing the largest number of Junior societies and Junior local unions, New Hampshire on securing subscriptions to *The Junior Christian Endeavor World*, and Connecticut on securing the largest number of Christian Endeavor Experts.

Iowa challenged all States that border on her territory on a 10-per-cent net increase in societies and a 20-per-cent net increase in new members. Nebraska accepted this challenge.

The Birmingham, Ala., union, with between 30 and 40 societies, challenged any union of the same size to a contest for organizing new societies. Tampa, Fla., and Province, R. I., both accepted this challenge.

Maine challenged New Hampshire and Vermont to unite with Maine in employing a field-secretary, and offered a contest in attaining the highest rating in an individuel society on the Efficiency Chart.

Texas challenged Illinois on having the largest number of registrations for the next State convention, and Illinois accepted the challenge.

Ohio challenged any State on the organization of new societies, Junior, Intermediate, and Young People's, gross count, on a two-year basis; challenged also any State on organizing Alumni Associations; challenged any State on missionary work—first, the organization of 300 mission-study classes; second, the enrollment of 3,000 persons in these classes; third, gifts for missionary work on a one-year basis.

Ohio also challenged Illinois, Pennsylvania, California, and New York to the enrollment of Comrades of the Quiet Hour in two years. New York and Pennsylvania accepted this challenge. Ohio challenged Pennsylvania to oragnize a purity department, and the challenge was accepted.

Pennsylvania challenged Ohio as to registration of delegates to the 1920 State convention, and Ohio accepted on a mileage basis. Pennsylvania also challenged Ohio in regard to the number of subscriptions to *The Christian Endeavor World,* and Ohio accepted on terms to be arranged later. Thus ended this valiant series of challenges.

"The Christian Endeavor Alumni Association and Its Possibilities" was the subject, an important one, assigned to Mr. Fred Ball of Cleveland, O. He was introduced by Dr. Poling, who spoke of the relation of the Christian Endeavor Alumni to the world-wide Christian Endeavor programme of reconstruction.

Mr. Ball told briefly and effectively the story of the organization in Cleveland. In that organization Dr. Poling was the moving spirit. In 1916 he visited Cleveland and suggested an organization that would tie up graduate Endeavorers to the cause of Christian Endeavor and assist in a financial way the programme of the United Society. At the first meeting 76 members joined the Alumni fellowship, and pledged more than $450 a year on a five-year basis to advance Christian Endeavor in all the world. Since this meeting the organization has steadily grown. Its membership has reached 467, and only 8 of this number have withdrawn from its membership. The pledges to-day exceed $2,350 a year for five years, and this sum is sure to grow.

This Cleveland Alumni Association is the pioneer Alumni Association in the United States, and Mr. Ball, its first president, has large plans to reach friends who are interested in helping the young people of the world. The Association has also the honor of having made possible the two journeys which Dr. Poling has made to Europe.

Alumni work is missionary work, and Alumni gifts go to missions. Ninety-nine members of the Association have increased their gifts to the Association because they know where every penny of their money goes.

The facts presented by Mr. Ball evidently gripped the delegates, and brought a vision of great things within the reach of most city unions.

Dr. Poling explained the Alumni plan, and announced that Mr. Stanley B. Vandersall will now give his whole time to the Alumni idea, and an *Alumni Quarterly* will be sent to every Christian Endeavor alumnus in the United States. The plan will indubitably enable the State unions to double the amount given by the societies for State field-work.

The closing address was given by Mr. David McConaughy, author of "Money, the Acid Test," and a leader in the Presbyterian New Era Movement, whose topic was "Stewardship, the Next Great Revival." The speaker pointed out that the stewardship revival is not a future thing, it is upon us *now*. The war has shaken us out of the comfortable illusion that men are the *owners of things*. They are not. The Government has large claims on the individual's income. *And so has God.*

Stewardship includes much more than money. Spending and saving are as much parts of stewardship as giving; in fact, stewardship must take in the entire scope of our life.

Stewardship is not a man-made scheme for raising money, but one of God's schools for raising men. Mr. McConaughy had with him a series of small charts to visualize the truth he was trying to teach. The whole world is a school and a workshop, and the marvelous thing is that while we use God's tools in His workshop we not only fashion things, but the things we fashion *fashion us*. Money is one of these things. The use of it fashions us in the realm of character, the most important thing on earth.

Mr. MeConaughy pressed upon the mind the need of every young person's making a definite stand, in the home, and in the church, and in the Young People's society, on the matter of giving. A standard must be set and lived up to. We must get into definite partnership with God and keep an account with Him. We must have method in our dealings with God. Most people who keep no account of their gifts to God think that they are giving more than they really give. We must know that when we are faithful in our dealing with Him, He will be faithful and generous in His dealings with us.

Stewardship, usually considered a "dry" subject, became, as the speaker talked, a most interesting and tremendously important theme. The audience was made to feel that the common sentimental view of Christian giving is totally and woefully inadequate. Certainly many got a new idea of their responsibility for the use of both money and life.

THE CHALLENGE CONFERENCE.

NORTH PRESBYTERIAN CHURCH.

Thursday Morning, August 7.

Other Challenges and Purposes.

The Endeavorers, large numbers of them, who went to the North Presbyterian Church on Thursday morning, at the time of the Quiet Hour, enjoyed a wonderful treat in hearing that truly great missionary, Dr. Robert Hume of India. He told how the missionary introduces a Hindoo to God. It was the gospel in simple terms, vivid, picturesque, as a wise father talks to his child. It got down to the fundamentals of the Christian faith. It made prayer very real, not only talking to God, but hearing God talk to us. It made missions and evangelism very real also, and was a splendid illustration of what can be done anywhere to open men's hearts to the Father.

Clarence Hamilton, field representative of *The Christian Endeavor World*, was the first speaker at Secretary Shaw's Purpose Conference. He had just returned from his parish in France of 300,000 doughboys, and he quoted a sentence of John R. Mott's which he heard "over there": "An act is never complete until it is recorded." He spoke finely of the work which our paper is doing to record the great work of Christian Endeavor, and of the necessity that this record shall be widely read if it is to do its full work.

President Wright of the Maryland union, introduced his conference by emphasizing the importance of the financial side of religion. His first speaker was Mr. Stanley B. Vandersall, national Alumni superintendent, who advocated strongly the adoption of the principle of the tithe by our Christian Endeavor unions. If the tithe is helpful to the individual Christian,—which is abundantly proved,—it is equally inspiring to a Christian Endeavor union. No union can do its best work without a budget, and the tithe should be in the budget from the start. This tithe is to be used for financing the extension of Christian Endeavor in all lands through the World's Christian Endeavor Union. Sixty-nine unions have already begun this plan; Kansas, for example, gives $500 a year out of its $5,000 budget.

The interest aroused by this proposition was evinced by the numerous questions asked Mr. Vandersall in the open parliament which followed. Equal interest was shown in Mr. Ball's presentation of the Alumni proposition, which followed the same lines as his presentation in the First Baptist Church.

One of the most hilarious and meaningful sessions in the history of Christian Endeavor was the challenge session then conducted by Secretary Shaw.

The District of Columbia opened the game by issuing a challenge to a 25-per-cent net gain in societies, and Maryland promptly accepted.

New York challenged California to a 50-per-cent increase in membership. Paul Brown, coming in later (detained 48 hours on the railroad), gladly accepted the challenge.

Bridgeport, Conn., challenged the whole State to a 50-per-cent membership increase, and the State local unions (most of them were represented in the Conference) accepted with alacrity.

Connecticut also challenged itself to carry out all Dr. Clark's challenges. Besides, it challenges Massachusetts, and every other State that does not support a field-secretary, to employ one for at least three months, at a cost of at least $1,000, between now and next summer. Massachusetts, which has already obtained Gates for six months, took up the challenge.

Indiana challenged Michigan on the one hand to organize more new Junior societies if it could, and Missouri on the other hand to organize more new Young People's societies. Michigan came back with a definite challenge to organize one hundred new Junior societies in two years, and Indiana accepted.

Missouri took up Indiana's dare, adding a challenge for the largest number of Efficiency Charts in use; and Indiana was not slow in meeting Missouri.

West Virginia challenged Oregon on a proportionate increase in *Christian Endeavor World* subscribers, which Oregon accepted, also taking up all of Dr. Clark's challenges. West Virginia also challenged Oklahoma on the number of Expert graduates.

Oklahoma raised that gage of battle, and proceeded to challenge Kansas for the largest number of new societies in two years. "We'll take you," said Kansas, and came back at Oklahoma by daring the State to duplicate the Kansas budget of $15,000 for three years; but Oklahoma wanted to think that over a while.

Illinois and Indiana combined to challenge New York State on the largest net increase in *Christian Endeavor World* subscriptions by March 1. New York promptly accepted and promised to deliver the largest number.

Detroit challenged the cities of Michigan to be the first to get fifty new Experts. Grand Rapids at once took up the challenge—bedsteads against Fords.

Tennessee challenged New Jersey on the largest actual and proportionate numbers of new Christian Endeavor Experts, and the dare was taken at once.

Louisiana challenged Georgia on *Christian Endeavor World* subscriptions, and Georgia accepted on the proportionate basis.

Baltimore met the District of Columbia's dare of the largest number of new Intermediate societies, and also Brooklyn's challenge on the whole two-year campaign.

One of the most interesting challenges was that of Connecticut, which "stumped" New Jersey to pass the Eighteenth Amendment first. New Jersey accepted, and Rhode Island—in the other church at the time—will doubtless get into the three-cornered contest.

The Disciples challenged the Presbyterians on the largest aggregate attendance at State conventions, and the Southern Presbyterians immediately accepted.

The eleven Southern States challenged the new Southwestern Federation to the largest increase in *Christian Endeavor World* subscriptions in proportion to the number of societies. Oklahoma and Arizona agreed, but Texas and the other States were in the other church.

Virginia's 262 societies issued a dozen challenges. Oklahoma met them on *Christian Endeavor World* subscriptions, West Virginia on new societies, Tennessee on *Dixie Endeavorer* subscriptions, and Maryland on contributions to denominational missions.

Michigan issued a general challenge to a 100-per-cent increase in Intermediate societies. This challenge was met by Connecticut, Oklahoma, Virginia, Tennessee, New York, Louisiana, Missouri, and the District of Columbia.

Idaho challenged Montana on Junior and Intermediate societies, and to have a Christian Endeavor society in every church or know the reason why not.

Grand Rapids issued an interesting "dare" to any city to have the largest mileage, in proportion to its population, in attendance upon the next International Convention. Bridgeport accepted. So did Fort Wayne, Ind.

Every summer New York City conducts 1,000 open-air meetings manned by Christian Endeavorers, and challenged any city to make a record "anywhere near it." New Britain, Conn., took up the gage on a proportionate basis.

Georgia challenged the Southwestern Federation and all Dixie to the greatest proportionate number of societies following the monthly service programme. Virginia and Tennessee at once accepted.

Florida issued an unaccepted challenge for the largest number of 300-per-cent societies.

Kansas took up the challenge of Missouri for the largest number of new Expert. Oklahoma met Missouri in its challenge for the largest number of subscribers to *The Christian Endeavor World* and the State Christian Endeavor paper.

South Carolina challenged Georgia for new members of the Quiet Hour and the Tenth Legion.

Finally, New York challenged any State for the largest number of Alumni Associations formed during the two years. This dare was taken, on a proportional basis, by Michigan, and also by California.

The closing address of this superb session was worthy of it. The theme was "Stewardship, the Next Great Revival," (For this Address in full see Chapter of Addresses), and the speaker was Rev. Joseph B. Baker, D.D., the eminent Lutheran pastor from Gettysburg, Penn. Dr. Baker is a master of pithy and pointed sentences, and his address was a wonderful combination of condensed wit and wisdom.

"Stewardship," he said, "is not generosity, for that thinks too much of itself. Stewardship is not charity, for that gives because it sees while stewardship gives because it ought. Stewardship is not taking up a collection, for that savors too much of the garbage-can. It is not a 'drive,' for that smacks too much of the whip, and Christ says, 'Follow Me.'

"Stewardship is administration. It means that what you have you have for another. An administrator under a will never thinks of sacrifice when he gives a check to an heir. Before the war we thought that all we had was our own. We were soon driven out of that thought, and made to feel that what we had we had for the Government first and for ourselves afterwards.

"Large recent gains in the receipts of mission boards show that the idea of stewardship is percolating through the minds of Christians. This is prophecy of a new revival of giving, which is greatly needed. Gladstone once said that the British navy could carry a message to every person in the world in six weeks. We have been twenty centuries about the task, and have not not yet carried the message of Christianity to all the world.

"This work of financing the Christian warfare is not the task of the leaders only. Before the war we thought that the work of financing the nation belonged only to the big bankers, but soon we found Uncle Sam asking the quarters and dimes of the boys and girls.

"The results of this revival of stewardship will be another flood—not a destructive one, but the flood of blessing of which the prophet spoke. It will add pentecostal power to the Church, for when Christians begin to part with what is closest to them, their money, sinners will begin to part with what is closest to them—their sins."

With this thought message a very powerful meeting came finely to a close.

CHAPTER VII.

FIELD SECRETARIES IN SESSION.

First Baptist Church.

Thursday Afternoon, August 7.

AS always, the meeting of the International Christian Endeavor Field-Workers Union was one of the tip-top meetings of the Conference. The president of the union, E. P. Gates, led the session in his breezy and inspiring way. The song service, led by R. A. Walker, was a model of how to do it.

After a prayer by Dr. Clark came reports from officers showing splendid work done by the field-secretaries to promote all the special kinds of Christian Endeavor work.

The field-secretaries spent much of their time in discussing in a fine spirit and very thoroughly Dr. Clarke's programme for the next two years' work suggesting a few modifications, and telling one another how they proposed to carry out the plans. It was a meaningful hour, full of practical purpose. It is the field-secretaries who will do a large part of the work in carrying out these large plans. Their hearts are in the programme. They will "put it across."

It was a series of rapid-fire talks. Superintendent Paul Brown told how to increase society membership. Secretary Lehmann told how to push denominational missions. Secretary Spafford told how to increase the number of societies. Secretary Breg told how to help the denominational leaders. Superintendent Vandersall told how to establish an Alumni Association in every union. Dr. Poling told of the United Society's new plans for aiding the work of State unions.

The monthly service themes suggested by the Executive Committee of the United Society were carefully discussed and adopted, as were the plans for Christian Endeavor Week for the next two years.

Then came some brief talks packed full of practical hints for the workers present in so large numbers. "How I Handle a Junior Meeting" was brightly told, with many illustrations and the exhibition of fine apparatus, by Miss Tschantz and Miss Hooper.

Mr. Reichel of Ohio, told how he handles an Intermediate meeting by placing responsibility upon the shoulders of the boys and girls, and doing as little as possible of the work himself.

Secretary Lehmann gave a very bright watch-talk to show how he would introduce Christian Endeavor to a lot of Sunday-school children. For instance, his watch had a gold-filled case, but Christian Endeavor is to make boys and girls gold all the way through.

Secretary Breg described the way to use summer institutes to advance Christian Endeavor. We need a delegated body, we need to knit them together after they arrive, we need to insist on study period, we must emphasize Christian Endeavor methods and allied themes, insert denominational hours, give over the afternoons to recreations and the evenings to platform meetings, including a "stunt night."

Secretary Spafford told how to build up rural societies by setting them to serve their communities, and gave a splendid illustration of a dying country which was grandly vivified through doing large work for its town. (See Chapter of Addresses.)

Mr. Shartle gave some wise suggestions on the correlation of the work of Christian Endeavor and of the denominations (See Chapter of Addresses), and Mr. Hamilton urged that in the State challenges *The Christian Endeavor World* should not be forgotten, but that challenges should be made for the largest number of new subscriptions.

Mr. Vandersall, speaking about the conduct of decision services, declared that they cannot be the product of mere arrangement, but must grow out of spiritual meetings. The character of the decision sought for must be made very clear and definite. It is better to have few or no decisions rather than have so-called decisions that are insincere or mistaken. We are not to look for numbers or records, but trust solely to the Holy Spirit.

"The Spiritual Message More Vital than Ever" was Paul Brown's characteristic theme. "If Christian Endeavor is anything," he said, "it is religious. We cannot beat other organizations to it in matters that are athletic, literary, and so on; but our task is to promote vital religion." He spoke very earnestly about the popularity of the dance and cigarettes promoted by the war, and insisted that Christian Endeavor and these harmful things cannot mix.

After a period of brisk and pointed discussion of the present condition of the spiritual life of the young people of our churches,—a discussion in which Dr. Poling took spirited and hopeful part,—this splendid and vital session was brought reluctantly to an end.

Part of the Audience at One of the Conferences. First Baptist Church.

Part of the Audience at One of the Conferences. First Baptist Church.

CHAPTER VIII.

CONSTRUCTIVE CONFERENCES.

FIRST BAPTIST CHURCH—CONFERENCE SECTION I.

Wednesday Morning, August 5.

Unions and Departments.

SECRETARY SHAW introduced with a few words of hearty appreciation W. Roy Breg, the secretary of the new Southwestern Christian Endeavor Federation, who led with great ability a live conference on the work of Christian Endeavor unions and departments.

That splendid worker and speaker, E. P. Gates, the newly-appointed field-secretary of the United Society, talked about ways in which our unions may increase their efficiency. First, the unions are to build up the local societies, for the union exists for the societies, not the societies for the union. Second, the unions are to work for the extension of Christian Endeavor. Third, the unions are to lead in many union efforts of Christians, for Christian Endeavor has been practicing denominational union for forty years while others have only been talking about it. This sprightly address hit dozens of nails right on the head and drove them home.

Paul Brown's train from California was many hours late, but his place to speak for the Intermediate work was effectively taken by Intermediate Superintendent Reichel of Ohio. He showed most effectively the great need of work for the Intermediate age, and pleaded with the Endeavorers to get that work on their hearts.

Miss Hooper spoke for the Junior department, and no one could do it better. "Yes you can make of the boys and girls anything you want to," she earnestly declared, "if you just want to hard enough. It is a splendid thing to *re*form, but it is ever so much better to *form*." She spoke of the tremendous need of trained leaders in all fields of service, a need that Junior Endeavor can meet, and she urged the Endeavorer to invest their lives right there, where they will count for the most.

Then, after some practical words from Mr. Shartle about Christian Endeavor literature and the Conference literature table, Chairman Breg conducted a capital open parliament on union work. The new Ohio Service Department was mentioned and described. The fact was emphasized that our unions are the key to the Christian Endeavor situation,—as they are strong Christian Endeavor is strong. The unions will not be strong unless the strongest leaders are selected. Our Christian Endeavor unions must emphasize Christian Endeavor above everything else —that way lies enthusiasm and success. Let us never talk about "dead" societies and unions. If their is one live member in a society, that society is not dead; but if there is one live society in a union, that union is not dead.

Denominational and Interdenominational.

That live wire, Field-Secretary Farrill of Wisconsin, led the Conference on coöperation between denominational plans and Christian Endeavor. The first speaker was Rev. O. T. Deever, the young peiple's secretary of the United Brethren. He related the happy experience of his denomination in passing from a strictly denominational to the broad Christian Endeavor platform in its young people's work. He told of the large gain in numbers and zeal made by his denomination when it appointed a young people's secretary who gives his whole time to promoting Christian Endeavor. And he rightfully insisted upon the need of coöperation on the part of the Endeavorer with all their denominational activities and aims.

The important topic of "The Sunday School and Christian Endeavor" was ably treated by Dr. Frank O. Erb, editor of the Baptist young people's and Sunday school publications. He illustrated the Sunday school and Christian Endeavor situation by the Siamese twins, whom he once saw. Each twin had a separate entity, but they were obliged to move together, or neither would get anywhere. He instanced the piano factory whose motto is, "If there is no harmony in the factory, there is no harmony in the piano." Dr. Erb pointed out the strength of each organization and the weakness of each, and the real need of each for the other. We have still to learn how to fuse the religious life of the young people of our churches with the life of the churches, and he described the council plan for coördinating the activities of the Sunday school, of Christian Endeavor, and of all other young people's organizations.

Many practical and stimulating points were brought out in Secretary Farrill's open parliament. The Endeavorers were most emphatic in their opinion that the Sunday school cannot take the place of Christian Endeavor. Dr. Stone gave very effectively his experience in his great church when he compelled all the Endeavorers above the age of twenty-one to leave the society and form a "Fellowship," the resulting society of really young people rapidly surpassing the Fellowship in numbers and in spiritual power.

That superb Chicago preacher and pastor, Dr. John Timothy Stone, treated in a masterful way the closing theme of the morning, "Spiritual Power and How to Gain It." "Do not confuse power with popularity," he said. "Many a man is so popular that he is powerless. Influence, too, is often not a matter of power. Spiritual power comes into a life when the Holy Spirit enters the life." He made very vivid to all of us the reality of the divine life in the heart of each true Christian. The Spirit opens to us the great meanings of the Bible, and gives us Scriptural stability and power. The Spirit comes to us in prayer. We cannot do anything worth while unless we pray. Nor can we have the power of God's Spirit unless we use the power that God has given us. If we have not hunger for the souls of men, something is wrong with us. God's Spirit comes to us as we identify ourselves with the problems of the place where we live, and put God into the solution of them.

NORTH PRESBYTERIAN CHURCH—SECTION 2.

Wednesday Morning, August 6.

The second section of the Wednesday morning conference nearly filled the beautiful North Presbyterian Church, with Rev. Morris E. Alling, president of the Connecticut union, as leader. The first topic here also was coöperation between the denominational and interdenominational agencies.

Mr. Alling pointed out that the Spirit of God depends to-day, as always, on consecrated personality. He made it plain that the church must have the voice of the Son of God, and provide for the Church of His love an earnest, consecrated leadership.

Secretary Deever spoke here also. Christian Endeavor, he said, has a wonderful name; it deals with life at a wonderful age; it has a wonderful message; and it urges wonderful basic principles that are fundamental to young life. One of the most wonderful things about Christian Endeavor is its ability to adjust itself to all denominational plans.

Dr. Deever pointed out the value of the Christian Endeavor Convention, State and International, as a promoter of Christian fellowship and union. At the same time the denomination can hold a network of denominational conventions for the presentation of matters of vital interest to the church.

Dr. Erb spoke also at this conference. "This is not a time," he said, "in which any organization should say, "I am superior, you are inferior; I am primary, you are secondary." In some respects all are primary and in other respects all secondary. There must be no talk of one organization's swallowing another. The functions of each are too different. The fish that swallowed

Jonah found that it had swallowed a worth-while man, *and had to give him up*. If one organization in the church swallows another, the result will be the same. We shall only have the trouble put upon us of reorganizing the society that has disappeared."

These stirring talks were followed by an open parliament which developed into a discussion of the topic on which Dr. Erb spoke. Dr. William Shaw answered some questions with eloquence and force. To train in ability you must put responsibility on the young people. He combated the statement that has often been made that the Young People's societies have had their day. He made it plain that Christian Endeavor has no spirit of antagonism or criticism with regard to the Sunday school. All that is asked is coöperation, each organization functioning according to its nature. It will be a bad day for the church if the young people are made to fight for their lives. It is wholly unnecessary, and Dr. Shaw said that he would do all in his power to avert such a calamity.

Mr. H. N. Lathrop, clerk of the United Society, was the leader of the conference on Christian Endeavor unions and departments.

The time spent in the open parliament proved exceedingly helpful, and dealt with the problems of both church and society. One speaker suggested occasionally throwing on the young people the entire responsibility for the evening church service. Another, a Wisconsin Endeavorer, told of his society's being the only religious organization in its village. The difference between the Young People's and the Senior Christian Endeavor society was emphasized. The Senior society usually consists of graduate Endeavorers, and the members of this society usually take for their meeting the midweek prayer meeting of the church, performing in that meeting the work which formerly they did in the Young People's society.

The session closed under the leadership of Rev. W. H. Hopper, D.D., president of the Kentucky union. The speaker was Rev. Floyd W. Tomkins, S.T.D., of Philadelphia, whose subject was "Spiritual Power and How to Gain It." Dr. Tomkins, known and loved far and wide as a spiritual leader in Christian Endeavor, took for the starting-point of his address the prophecy of Joel regarding the coming of the Spirit. Spiritual power, said Dr. Tomkins, is the supreme power. We shall never get it if we think of it merely as *a power* among powers and not the power supreme. Physical power has its place, and so has mental power, but neither can take the place of God's Holy Spirit.

Spiritual power is not the power that enables us to move men; it is the power that brings *us* to God. By it we enter into fellowship with Him, so that God is real to us. It comes to us through association with Him; we are partners with Him and work with Him.

How can we gain this power?

We must have simple and absolute faith in God, and because we have faith in God, also faith in men and in what God is doing among men. Power increases, too, as we use it. Do not wait for full power; use what you have, however little, and that little will grow. But the use of this power must be unselfish service for others. We are saved to serve. We must keep our ideals high.

FIRST BAPTIST CHURCH.

Friday Morning, August 8.

Life-Work.

"One of the greatest problems before the church of to-day," said Secretary Shaw at the Decision Service, "is the problem of workers. Money is given to-day for the Kingdom far more largely than ever before, but money is useless without men."

Dr. J. Campbell White, secretary of the Life-Work Department of the Interchurch World Movement, the speaker of the hour, emphasized Dr. Shaw's statement. Vocational stimulus and a guidance in secular callings are now given freely by our Government, and the church must do the same for religious vocations. The Christian Endevor societies and the Sunday schools can do much in this direction.

First, let all of us realize that God has for each one of us a plan related to His great plan for the world,—all our work prepared for us before we touch it. Second, let us all set our minds on finding what God's plan for us is, and then follow it wherever it leads us.

There is urgent present need of at least 40,000 more workers abroad and an equal number at home. These all need the very best possible training, an education that includes religion. No education can be comprehensive that eliminates God and the truths that He has revealed.

"God will speak to you when you are nearest to Him," Dr. White urged. "Do not be afraid of the higest spiritual impulses that He may send to you."

After silent prayer Dr. Shaw asked all Life-Work Recruits, Student Volunteers, ministers, missionaries, and other full-time Christian workers to rise. Then he read the covenant of Christian Endeavor Life-Work Recruits, and asked those who wished to take it to come to the front of the room. At once eleven young men and women, without urging, came quietly forward, and Dr. Tomkins made a tenderly beautiful prayer of consecration. Thus ended one of the most helpful and profitable services of Convention week.

First Baptist Church—Section I.

Friday Morning, August 8.

A Successful Society.

Amos R. Wells, in opening the conference on a successful society, said that such a society is like a successful man. A man, to be successful, need not be rich or famous or powerful; he need only follow Christ and do His will. So a society, to be successful, need not be large or widely influential; it may be very small and obscure. But it will be a successful society if it is doing, in Christ's spirit, the work that Christ is asking it to do. The test of a society's success is its Alumni.

Four factors of a successful society were treated by experts. The beloved Dr. John F. Cowan treated "A Good Prayer Meeting." "Buy it," he said. "Buy it with hard work. The price of any victory is blood, spent if not spilled. You will leave nothing to chance, if you want a good prayer meeting. A good salesman has regard to every least detail, even to the buttons on his coat. A good prayer meeting is good salesmanship for Christ."

President Wyatt A. Taylor of the South Carolina union, gave a fine description of a good social. He urged four particulars: (1) It must have a plan for starting, such as giving each man a slip of paper with the name of a State on it, and the young women slips bearing the names of State capitals, each State to find its capital; (2) it must give everybody a part; (3) it must have a programme; (4) it must contain some religious feature, opening or closing with song and prayer.

"A Good Missionary Meeting" was brightly described by Miss Ida C. Clothier, Colorado's field-secretary and missionary superintendent. It must not be all *miss,* but it must get as many young men into it as possible. A definite part in the meeting must be given to every member of the scoiety. Every Endeavorer should be in personal touch with at least one missionary worker. All parts of the meaning should be focused upon one central theme. No one should read from book or paper, but all should speak out of full minds and hearts. Christian Endeavor missions means a sympathetic outlook, in living service, over all the universe of God.

Miss McFee, president of the Rhode Island union, made many strong points regarding committee work. Committee service is of value because it deepens the knowledge of God. The committees should be carefully selected by a nominating commitee long in advance. The pastor and president should meet with the committee as often as possible. Some committee rules: (1) Know your job; (2) plan your work; (3) work your plan; (4) stay through to the end; (5) and everywhere and through it all, *prayer.*

The open parliament that followed was a lively one; many questions were asked and many experiences were given which will bear rich fruit in coming years.

Citizenship Plans.

Treasurer Shartle opened his citizenship conference with a little patriotic talk, and then introduced Dr. J. Stanley Brown, secretary of the Savings Division of the United States Treasury Department. The war, he said, taught us that we were the most thriftless and wasteful nation in the world. Only three millions out of our one hundred and ten millions had any savings-bank account; now more than thirty millions have taken up Government securities.

Half of Christ's parables have to do with money and thrift. Our Government is Christian in asking the churches to emphasize thrift. Educational pamphlets have been prepared by the Government, and may be secured free through the Savings Director of your Federal Reserve District. Let the Christian Endeavor societies form Thrift Clubs everywhere, for the broadening of vision and the stabilizing of finances. Financial morale will act as the antidote to financial panic, and it is not foreign to religious morale, though the latter is not to be measured in money terms.

President Coffin of the New York State union, discussed the Americanization of foreign-born peoples. This problem has grown out of our own neglect. It is a national problem with a community solution.

A man is not an American just because he was born in America; he is a man in whom America has been born. Americanism means liberty to promote the welfare of the whole people. It means a democracy which involves equality and coöperation. It stands for religion; ours is a Declaration of Dependence, dependence upon God. It stands for personal activity. It is one of the curses of American life that the ordinary American can be so complacent with evil conditions as long as they do not touch him personally. Dr. Coffin closed with many practical suggestions for this great work of Americanization.

Field-Secretary Hetzler of West Virginia, treated the important theme of community service. The world is fast becoming a vast community, but he who aspires to world service must prove himself in the service of local communities.

Community evangelism is one of the most challenging of community services. If the church seeks men, men will believe in the church. A wide plan of visitation is practicable for Endeavorers, together with the making of a religious census. The Endeavorers may start the agitation for reforms, promote improved social life, beautify community life by the introduction of music and art, flowers, lawns, and parks.

Mr. Paul V. Shaw then spoke of Y. M. C. A. work for foreigners. (See Chapter of Addresses.) Our country has 10,000 foreign-born students; the eyes of the world are upon us. Christian Endeavorers may greatly help by inviting these foreign students to their socials, to their church meetings, to their homes. Mr. Shaw gave an example of the way a little kindness shown such young men brings large results.

After a lively word from Hamilton on *The Christian Endeavor World* and some more brisk challenges to service, together with a brief but pointed open parliament, President Bond of West Virginia, introduced the closing period of the morning. The address was by that master of public speech, Dr. Ira Landrith, the subject being "What the Community Has a Right to Expect from the Church."

First, Dr. Landrith deprecated silly criticism of the churches and pastors. "The church of God," he said, "at its weakest and worst is better than any other organization on earth at its strongest and best." He paid a magnificent tribute to the average American church and pastor.

He paid his respects to the man who declares that "the business of the church is the salvation of souls," but hasn't brought a soul to Christ for the last twenty years. The main business of the church is the saving of souls, but whose business is it to destroy what is destroying souls? Christ came that men might have life and have it abundantly—abundant social life, commercial life, political life, such a life as Christ would approve and work through. The church of God must create an atmosphere in which no injustice or unrighteousness can live.

We must get rid of race prejudices, even prejudice against Germans who have been true to America; rid of sectarian prejudices, for if we are going to mix with Christians of other kinds in heaven, we must learn to get acquainted with them on earth; we must get rid of sectionalism and partisanship, if we would build up a true nation.

North Presbyterian Church—Section 2.

Friday Morning, August 8.

Society Success and Citizenship.

The services in the North Presbyterian Church on Friday morning began with a half hour's talk by Dr. James H. Pettee, president of the Japan Christian Endeavor union, who combined a beautiful Quiet Hour's service with a fine talk on "Hands Across the Sea." Christian Endeavor, he said, is the hand of the church across the sea. The hand, used in salutation, is a beautiful sign of friendship, and it is one which was invented, not in the East, but in the West. The Japanese and other nations bow, or kneel, but no Oriental nation has the mutual handshake. This method of friendly expression is Christian. It is the friendly touch. The hand is also used with which to beckon. But no heathen nation ever at first asked for foreign missionaries to come over and help them. The initiative has always been in the hands of Christians. Christ said "Go," for He knew that the world would not at first say "Come."

Southern-States Secretary Lehmann was the stimulating leader of the first conference. The topic was, "How to Have a Successful Society."

Field-Secretary Spafford of Michigan, briefly and clearly told how to have a good prayer meeting. The essentials, he said, are at least three: (1) Prayer by the leaders, and by others that the meeting may be a blessing. This includes the pre-prayer service. (2) Planning. This involves hard work for the prayer meeting committee. (3) Purpose. We must define our purpose and prepare plans to attain it. Mr. Spafford added a large number of practical suggestions such as advertising and sending invitations to the Bible school. Pray earnestly, plan faithfully, and define your purpose, and the meetings will be successful.

Miss Vida Leamer, field-secretary of the Nebraska union, spoke on "A Good Social." She used the letters of the word "social" to describe the successful social time: "S" means "start on time." Be social. Become acquainted. Shake hands. Start something in which every one may take part—kindergarten games, if necessary. "O" stands for "offering a prepared programme." Socials must be planned. "C" stands for "Christian Endeavor." The social must emphasize Christian Endeavor in some way. Work it into the plan. "I" stands for the "introduction of new games." "A" stands for "all together," all working like a team for a common end. "L" means that we "look out for new members."

"A Good Missionary Meeting" was the topic assigned to Rev. L. C. Little, field-secretary of the Louisiana union. Mr. Little, who is a Methodist Protestant, an "M. P.," said that he hoped that the letters in his case would also mean Missionary Propagandist. He spoke most enthusiastically of missionary work, and pointed out the necessity of planning the meeting and putting care and thought and love into its preparation.

"Good Committees" was treated by Miss Mary B. McPherrin, field-secretary of the Iowa union. Committees must be committees in more than name. They must be organized for service, and give to every member his work. Care should be exercised in the selection of members of committees. Remember, we are training leaders. Committees should change. Some are reorganized every six months, others every twelve months. Miss McPherrin earnestly urged that committee members secure information, through Christian Endeavor helps, about the work of their committees. She made it clear that good committees must plan their work in regular committee meetings.

In the open parliament that followed, a poster committee was suggested whose work would be to get out a poster once a week, once a month, or as often as possible. The plan is largely followed, for many Endeavorers described many posters used. Poster contests were also helpfully suggested.

Charles F. Evans, general field-secretary of the All-South Extension Committee, was the leader of the second part of the conference.

Mr. Paul V. Shaw made here also a plea for work with foreign students. If Endeavorers wish to know how to help foreign students, they should write to Paul V. Shaw, director Latin American Division of the Committee on Friendly Relations among Foreign Students, 347 Madison Avenue, New York City.

"Enforcing the Eighteenth Amendment" was the subject assigned to Rev. Morris E. Alling, president of the Connecticut Christian Endeavor union. He denied the statement often made that prohibition cannot be enforced, and stirringly answered the arguments political and economical, advanced against the abolition of the saloon.

Rev. H. F. Shupe, D.D., editor of *The Watchword,* admirably presented the topic, "Americanization of Foreign-born Peoples." America, he said, is a melting-pot, but we need a finer screen than we now have that will keep out undesirable citizens. He argued for the admission annually from each land of about five per cent of the number from that land already living here. He called for a Congressional Immigration Commission whose work would be to advance Americanization of foreigners through educational campaigns in public schools and factories. But the church's great task is to show the Christian meaning of the word Americanization, and through all sorts of service, visiting, supporting district nurses, and so forth, to manifest the spirit of Christ. In this task Christian Endeavor may have a wonderful share.

Dr. J. Stanley Brown brought to this conference also his message on Christian Endeavor Thrift Clubs.

Dr. Poling, leading the open parliament, outlined briefly the citizenship programme of the United Society. It will deal with thrift. It will deal with the enforcement of prohibition. It will deal with the duties of the individual as a citizen, the duty of voting, for instance. Endeavorers must interest themselves in the programmes of accredited citizenship organizations. We must ask for these programmes. We must Americanize by teaching English to foreigners. We must make surveys of the community. On these points Dr. Poling gave many instructive illustrations drawn from the work of Christian Endeavor societies.

Rev. H. H. Price, D.D., president of the Nebraska union, presided over the closing part of this conference. He introduced

Rev. J. T. McCrory, D.D., moderator of the United Presbyterian General Assembly, whose subject was "What the Community Has a Right to Expect from the Church." This was an eloquent address which brought out the truth that the only way Christ can reach the world to-day is through His people. The Master has placed us where we are in order that through us He may save the world, and the world has a right to look for saving power in the church. If the salt has lost its savor the world has a right to repudiate the church.

While we talk about reconstructing this war-devastated earth, we must remember that God also has a plan of reconstruction, and His plan is to reconstruct the heart, regenerate the souls of sinners. When that is accomplished we shall indeed have reached the era of universal peace and brotherhood. In the meantime the church must go into the community and save the souls of men. If it does not do that, the church is failing. We must have the power to convert souls.

The community has a right to expect that the church inculcate good will among men. For this, Jesus came into the world. But the church must not only *teach* good will, it must *show* it in practical ways.

The world has a right to look to the church for education. The church has the truth about God, and its duty is to teach it.

The world has a right to look to the church to take care of the children, and we are not doing it well enough. In fact, we are losing our own children.

Finally, the world has a right to look to the church for the highest ideals of community life. Somebody has to lead the way in all departments of community life. This is the church's duty and task, a truly great and important task. The church must lead the world toward the millennium. There is no hope in any other direction than the church. If the church fails, the world fails. But it will not fail, however long it takes to attain its ideals. Compared with the world, the church is small; but it is armed for its task with the mighty, unconquerable power of God.

First Baptist Church—Section I.

Saturday Morning, August 9.

Standards

Under the skilful guidance of Dr. Gilbert Glass the Saturday-morning conference on the unifying of denominational, State, and United Society standards moved briskly and helpfully.

After the opening prayer of Amos R. Wells, the first speaker was Dr. Frank M. Sheldon, the experienced and able secretary of the Congregational Educational Society. He declared that Christian Endeavor leaders are right in thinking that the church is not enough interested in the young people's programme; but neither is the church sufficiently interested in the Sunday school programme. There are in this country thirty million young people under the age of twenty, and eighty per cent of them will become Christians, if at all, before their twentieth birthday.

There is need of unified work. There is far more work than either the Sunday school or the Christian Endeavor society can do alone; the church needs them both. The work of young people should be enlarged. They should be given larger responsibility and work in the church.

One of the finest things about Christian Endeavor is that it works through the church, it does not make of itself a separate denomination, it is adapted for whatever the denomination wants it to do.

We believing in grading our work to meet the different needs of unfolding lives, and in this Christian Endeavor can greatly help the denominations.

Secretary Shaw discussed United Society standards in his illuminating way. We can unify standards, he said, by simplification. Often our standards need too much bookkeeping. Three sets of standards, from the United Society, the State union, and the denomination, so confuse the societies that they give up in despair. We can clarify and classify our standards. We can coöperate. We can come close together, as Christian Endeavor gets people together. All we need in our work, in order to achieve what all of us want, is to be kindly considerate of one another and all pull together.

Field-Secretary Spafford of Michigan, spoke of the value of State standards. When standards are suggested for the nation, it is helpful to subdivide them into State goals; but they should all be harmonized; no confusing standards should be put forth by the State unions.

A lively and profitable discussion closed this conference. The second division of the morning was led by Secretary Breg of the Northwestern Federation, and continued the general theme of standards.

Dr. Clark introduced his presentation of the 1919-21 Christian Endeavor Standards with a few words showing the immense advantage of setting up goals. They make our purposes definite, and even if we did not reach the goals—but Christian Endeavor *does*—we should go much farther than without the goals.

In speaking of our purpose to add fifty per cent to the membership of our societies, Dr. Clark urged that we make far more than we have made of associate and coöperating members. As to increasing the number of societies, Dr. Clark urged fresh emphasis upon Junior and Intermediate societies. As to following up the denominational plans, Dr. Clark declared that we have the denominational leaders in this Conference, and in favor of

Christian Endeavor, more now than ever. As to placing field-secretaries in every State, Dr. Clark declared that the States that shine now are those that have employed full-time secretaries; and there is no financial problem, for the right field-secretary will solve all the financial problems of the State. Finally, as to the new Alumni movement, Dr. Clark believes that it will signalize this Conference more than anything else.

As Dr. Clark had to leave for another conference, the audience rose and with uplifted hands promised: "Dr. Clark, we'll see it through! Come on. Let's go."

Secretary Lehmann was introduced as the man who originated the Monthly Service Topics. He went through the programme for the next two years month by month, and told in his inimitable and arousing way how the programme would be carried out in Dixie. His talk bristled with good points, and pencils were kept busy.

Editorial Secretary Anderson discussed the Christian Endeavor Week, going through the plans for the various days and giving the reasons for each. He showed the value of celebrating church festivals such as this one. It is not a commemorating celebration, but is forward-looking, purposeful. Our paper will explain these plans in full as Christian Endeavor Week draws near.

Intermediate Superintendent Brown next outlined the Loyalty Campaign, which is his vivid and attractive setting forth of the plans for the coming two years placed before the Endeavorers by Dr. Clark. This formulation of our goals will become a classic, and we will bring it often before the readers of this paper.

The morning conference at the North Presbyterian Church on Saturday morning was admirably led by Rev. C. R. Sine, president of the Ohio union. The topic and speakers were the same as those in the First Baptist Church reported above, the speakers going from church to church.

The open parliament brought out the interesting fact that many societies that have tried out the Christian Endeavor Week programme in former years have found it a success. Only two societies admitted that the celebration of the Week had failed, and stated that the reason was weather conditions. One society told how it began by celebrating the two Sundays, and from this built up, in succeeding years, a whole week's programme. Another society was mentioned which found a new lease of life by celebrating the Week. Preparation, surveys, and prayer lists were suggested as means of making Christian Endeavor Week a source of blessing.

CHAPTER IX.

THE ALUMNI BANQUET.

ELMWOOD MUSIC HALL.

Thursday Evening, August 7.

THE Alumni banquet in Elmwood Music Hall on Thursday evening proved to be an occasion of abounding joy. Eight hundred and fifty young people filed to the tables and enjoyed—that is the word—a first-class dinner; enjoyed also, beforehand, and during dinner, the songs and the yells rendered by the various delegations, which were able to sit together. Vitality! No conventions are like Christian Endeavor conventions in this respect. The young people are carried forward on a mighty tide of *life* that *must* express itself. One realizes that this flood of enthusiasm will flow in town and village into channels of splendid service.

And what singing, led by an orchestra! Mr. Gowing, the choir leader, made the various delegations sing a convention chorus to the tune of "Smile awhile." And how they sang, and smiled, and sang yet more lustily! There was no need to try to put "pep" into this crowd. The "pep" was there all right, and in full measure.

Dr. Clark presided, and Dr. Poling proved himself an adept toastmaster, introducing the speakers with witty and fitting characterization and illustrating his thought with an inexhaustible fund of anecdotes that sent the delegates into gales of laughter.

Then there was a song to Dr. Clark which the delegates sang with ardent love to the tune of "Auld Lang Syne."

> Could we forget, could we forget
> The one we've known so long;
> Could we forget him, as we cheer and as we sing our song?
> Oh, Dr. Clark, we sing to you, our love we give you, too,
> For we are proud of our C. E. and owe it all to you.

Dr. Clark modestly disclaimed the honor of its all being due to him, but the delegates showed their opinion by repeating the song under a Chautauqua salute.

The topic, of course, was the Alumni movement, which Dr. Clark said is the most important movement that has arisen in Christian Endeavor in the last ten years. In a few words he elo-

THE ALUMNI BANQUET. ELMWOOD MUSIC HALL.

quently described the Alumni idea, which makes a permanent place in Christian Endeavor for all Endeavorers who grow too old for the society, or who, on account of other labors, must give up its activities. Just as college alumni are proud to help their *alma mater* in a financial way, so in days to come Christian Endeavor Alumni will be happy to share in putting through the world-programme of Christian Endeavor.

Dr. Poling truly characterized this meeting as the first national gathering of the Alumni of the United Society of Christian Endeavor, and he indicated that those present would have the opportunity of becoming charter members of the national Alumni organization. This was, as it were, the first national class reunion, and, of course, reminiscences were in order, and Dr. Poling indulged in them in his own enthralling and eloquent way. But reminiscence is not all. To-night the speaker said, we shall not look backward, but forward to the vast tasks of our Christian Endeavor society in the future.

Mr. Fred L. Ball, president of the Cleveland Alumni Association, the pioneer Alumni organization, was fittingly the first speaker. He told the story of the organizing of the Cleveland association and of how the association had just determined to finance the work of a Christian Endeavor secretary in France to help to rebuild Christian Endeavor in that stricken country. These Alumni are interested because a definite programme is put before them, and they know what is done with the money they give. Mr. Ball challenged the Endeavorers to take a forward step and start an Alumni fund of $100,000 to finance the work of Christian Endeavor.

Mr. Ball brought with him, and presented to Dr. Poling from the Cleveland association, five Liberty Bonds worth $250, and in addition a check for $750, making a gift of $1,000 for Christian Endeavor work. One can imagine the delight this happy surprise brought to the Endeavorers.

There was another surprise, this time from Amos R. Wells, who, less than half an hour before, had been called upon to produce an Alumni poem. Forthwith he brought forth the following "Alumni Ode"—only, as Christian Endeavor Alumni should not owe anything, he called it an

Alumni Paid.

(Dedicated, on request, to Hon. Daniel A. Poling, LL.D.)

The Christian Endeavor alumnus,
He sure can make a rumpus,
He yells and cheers
To split your ears
And you can bet it's some fuss.

The Christian Endeavor alumni
Find that their frolics come high.
Ten dimes a year,
Or more, I fear,
"Oh, how expensive!" some sigh.

But, alumni of Christian Endeavor,
We'll sing and cheer forever,
We'll yell and holler
And pay our dollar,
And never resign, oh, never.

But Drs. Clark and Poling had seriously objected to the last stanza. They were not satisfied with one dollar apiece. They were mercenary, said Wells, and he was literary. So, though the Muse seriously objected, he felt obliged to submit the following revised version:

The alumnus of Christian Endeavor,
He'll sing and cheer forever,
He yells and hollers
And pays his dollars,
And never resigns, no, never.

Dr. Poling called on Morris E. Walling of Connecticut, who had been leading his State delegation in some of the "peppiest" songs of the evening, to teach the audience how to do it, and, standing on the platform table, as agile as Billy Sunday, he made the people sing, "Pack up your troubles in your old kit bag," as surely they had never sung it before.

The next speaker was Carroll M. Wright of Baltimore, who took up the Alumni tale in reminiscent vein, which, however, soon changed toward the true sphere of Christian Endeavor, the future. In Maryland the Alumni are put to work. A definite sum has been pledged to the United Society. When it is necessary for Maryland's union officers to reach out-of-the-way places, they call on the Alumni to take them to their appointments in automobiles, and these Alumni not only lend their services in this way, but also take pleasure, for old times' sake, in addressing the Christian Endeavor rallies. Maryland is opening avenues of service for the members of its Alumni Association.

W. Roy Breg, new secretary of the Southwestern Christian Endeavor Federation, gave a vivid description of the extent of the territory he is now to serve. In all this vast territory to-day, there are fewer Young People's societies in Illinois alone; but a great Christian Endeavor development lies ahead in this region. He told how Texas Endeavorers, who have given subscriptions to the United Society, are eager to see the money used for the spread of Christian Endeavor in the republic of Mexico. Speed the day when the United Society shall be able to send a field-secretary to this great and needy country.

Miss Agnes Baker, New York's beloved Junior superintendent and member of the convention committee, called to the platform, broke into rhyme in a hearty word of greeting to the audience.

Treasurer Shartle told of a man who, standing at a small stream, was told that he was standing at the fountain-head of three great rivers; he drew the lesson that to-day we are at the fountain of the Alumni Association, and uttered a stirring call to "come on" and make the association a tremendous success.

R. A. Walker, formerly of Indiana, now of the Chicago office of the United Society of Christian Endeavor, declared that if, as Dr. Poling said, he has developed a smile, he has discovered it in the Intermediate society of Christian Endeavor. Mr. Walker told of his determination to try to make his work in Chicago a real service for Endeavorers in bringing printed helps to them to increase their knowledge and efficiency. Mr. Walker closed by getting the audience to join in singing a song beginning:

Dr. Clark, we're glad that you are here,
Every time you seem to grow more dear.

Miss Grace Hooper, national Junior superintendent, gave reasons why she believes in Junior Christian Endeavor: first, because she believes in God, and because there are boys and girls who need our society and who are ready to respond to the call. Junior Endeavor gets ten-per-cent service from the church, but it brings a ninety-per-cent result among the boys and girls. As long as there are children in our homes and in our churches, there will be reasons enough for the Junior society.

The field-secretary of the United Society, E. P. Gates, formerly of Illinois, received a hearty welcome and delivered a whirlwind address, which is his ordinary style. His heart warmed as he spoke of the Chicago Christian Endeavor union and told of some of the fine things it is doing; he prophesied that when Chicago really gets started on the Alumni movement, she will show that the last two letters in the name of the city still mean "GO."

An old-time friend of Endeavor, a trustee of the United Society, Dr. Samuel H. Woodrow of St. Louis, was next called upon to speak. Dr. Woodrow has been connected with Christian Endeavor since 1883, and is still as deeply interested in the movement as in those early days. He paid a warm tribute to Dr. Clark, and declared, out of twenty-five years of experience as pastor, that no organization has yet been devised that can take the place of the Christian Endeavor society. The church *must* have some organization that will do the work that Christian Endeavor is doing. The Sunday school has a fine task, but in it there is no real place for expressional religious activity. We must have training for the young, and we can get it adequately only in the Christian Endeavor society.

Then came Stanley B. Vandersall, now Alumni superintendent of the United Society. He earnestly announced that if there is any virtue or talent or ability in him, they will be wholly devoted to the advancement of Christian Endeavor. Mr. Vandersall briefly mentioned some of the Alumni possibilities, among them a great body of at least 5,000,000 former Endeavorers to whom appeal may be made for support for Christian Endeavor.

The audience rose to sing "Blest be the tie that binds" as a welcome to Dr. Robert A. Hume, the eminent missionary from India, a former president of the India Christian Endeavor union. He pointed out feelingly that the greatest thing that God can do for a human being is to make him love, and the way in which God makes people lovers is by making them endeavor to serve. This is the genius of Christian Endeavor. He stated that when the Society was organized Dr. Clark realized that the weakness of the young people is not of the heart, but of the will.

We know the way our feet should pass,
Across our hearts are written Thy decrees;
Yet now, O Lord, be merciful to grant us more than these.
Grant us *the will* to fashion as we know.
Give us the strength to labor as we feel.
Grant us the purpose ribbed and edged with steel
To strike the blow.
Knowledge we ask not, knowledge Thou hast lent.
But, Lord, *the will*, there lies our bitter need.
Grant us *the will* to build above the deep intent
The deed, the deed.

But if we need this training of the will here, how much more do we need it in lands beyond the sea where the people think that all religious work must be done by the priests! Christian Endeavor sets up nobler ideals, and shows our young folks in India that there is work for *all* to do.

For years the United Society has supplied the means for supporting a Christian Endeavor secretary in India. There is no Alumni Association there as yet, but Dr. Hume pledged himself to use every effort in his power to establish Alumni Associations in India.

Years ago a Hindoo policeman found a house that was suspiciously shut up; he opened the door and discovered the dead body of a woman with a baby trying to suck at her breast. The policeman took the child to a missionary, and to-day this boy is in training to be a Christian Endeavor secretary. Shall we not support him?

Dr. Poling then presented Mrs. Francis E. Clark, who received a magnificent ovation. Mrs. Clark was in her happiest vein, bright, sparkling, humorous. She claimed that she was the first alumna, having belonged to the first society. She referred to lessons in thrift which we have all been taught in recent years. These are days for economizing, and if we are careful with what we have, we shall be able to give generously to send Christian Endeavor secretaries to help push the foreign work of Christian Endeavor.

Then followed Dr. Ira Landrith, extension secretary of the United Society, who opened with a wit made pathetic by his having been forced, as he alleged, to listen to all his best stories from the lips of earlier speakers. He had just come from Butte, Mont., which gets its water from the other side of the divide where flows the Columbia River. After water has been taken from the river to meet Butte's needs there is still enough in the river to irrigate a thousand valleys. The graduate Endeavorers have a-plenty of this world's goods, and there is money enough, as there is water enough in the river, to meet the needs of the movement for the betterment of the world.

Dr. Landrith told with great fervor of the big things that Christian Endeavor is doing through men who are receiving quite inadequate remuneration. Christian Endeavor is a necessity, and will be as long as new babies are born. Nothing has taken its place, nothing can, and nothing ever will take its place in any adequate way.

The speaker said that if he has accomplished anything in citizenship or for prohibition, he owes it all to the inspiration he received in a Christian Endeavor convention in Tennessee twenty-five years ago. Dr. Landrith said that he was twenty-five years behind in his Alumni Association dues, and he promised to pay up by presenting to the organization his Liberty Bond.

General Secretary Shaw stood on the table to make the closing address in order to get higher than the well-named "extension" secretary, Dr. Landrith. In great pride he took the splendid audience as his theme and argument. The magnificent company of present-day Endeavorers proves the value of the Society. Dr. Shaw never went to college, but he is proud to be an alumnus of Christian Endeavor. He has given the space of a generation to Christian Endeavor service. He painted a stirring picture of the splendid growth of the movement which he has witnessed, and made a powerful plea for the fullest possible support of the United Society of Christian Endeavor and its splendid, world-wide work, backing up his words by the gift from himself and his wife, of one thousand dollars.

A meeting like this could end in only one way, namely, that of giving the young people an opportunity to follow Dr. Shaw's example in making some thanksgiving contribution to Christian Endeavor. One after another rose in rapid succession and pledged, one $50, another $25, another $25, another a Liberty Bond, and so on in genuine rapid-fire manner and with great enthusiasm. These were not Alumni contributions, but special thank-offerings, many of which were reckoned on the basis of a dollar for each year during which the donor had been connected with Christian Endeavor. When the pledges were all in it was found that this thank-offering amounted to the splendid sum of $4,637.

Cards were passed to a great crowd that stood and signified their readiness to make a gift representing at least $1 a year for each year during which the donors have been connected with the Society.

Then the Alumni cards were passed out, and those who desired to become members of the Alumni Association were given an opportunity to sign up. Besides the amount of the thank-offering, mentioned above, and Cleveland's $1,000, 265 Alumni who signed the cards pledged a total of $4,248 for five years. The total of gifts for the evening was thus $10,000.

This thrilling meeting, which lasted until well past eleven o'clock, was closed by Dr. Clark with grateful prayer and the benediction.

The Sunrise Prayer-Meeting. State Normal School Grounds.

CHAPTER X.

SUNRISE CONSECRATION SERVICE.

On the Grounds of the State Normal School.

Saturday Morning, August 9.

"THIS is one of the greatest surprises of my life," said Dr. Clark in opening the six o'clock (really *five* o'clock) open-air prayer meeting on Saturday morning. He had expected only a few, as the meeting of the preceding evening was so prolonged; but he saw before him more than 500 bright-faced young folks, eager to meet God "early in the morning."

The meeting was held on the spacious and beautiful grounds of the State Normal School, with stately buildings in the background. The benches were crowded, and many were standing.

Mr. Gowing led the singing, which was full of spirit. Dr. Clark called for Bible verses, and scores rose promptly to their feet, repeating the strong Scripture sentences in rich variety, many speaking at once in different parts of the audience.

After a moment of silence, Dr. Clark led us in an earnest prayer. Then he spoke of some of the notable sunrise prayer meetings that he had attended. He recalled the very first of them, held in connection with the third Convention, held in Saratoga. The janitor of the church could not believe that young folks would really come out at that unheard-of hour, so that when a great crowd of Endeavorers besieged the Convention church at sunrise they found the building locked and had to hunt up the janitor. Then he remembered a sunrise meeting in Kobe, Japan, held on a hill to which they had to ascend under many torii, those emblems of heathenism. Again, in Rio de Janeiro he attended a sunrise meeting on top of a mountain which rises from the heart of the great city, three thousand feet above the sea. During the meeting the sun came up around the corner of the mountain and struck full in their faces, and some one at once began to sing in Portuguese, "The morning light is breaking." "Thus in all the world," Dr. Clark concluded, "our Christian Endeavor brothers and sisters are praying with us, and we are in a mighty fellowship."

The principal message of the meeting was brought us by Rev. Dwight Goddard, the well-known writer of books on mystical religion. He gave us a most illuminating little talk on the distinctions between body and mind, and *sub*conscious mind and the *super*conscious soul. We accomplish much by the cultivation of body and mind, but the subconscious entity is indispensable, carrying on as it does most of the essential business of our living on the lower plane, such as breathing; and the superconscious entity is even more important, for it is in this sphere that man comes into vital contact with God. Few persons, comparatively, cultivate this superconscious entity; few persons, by quiet meditation and trustful communion, come to know God immediately and powerfully. This blessed energizing and purifying experience may be enjoyed by all that will, and the endeavor to enter into this mystical but very real knowledge of God is blessed and fruitful beyond measure.

We were next uplifted by a purposeful consecration service, led by Dr. Clark. After a fervent prayer by our leader, we all repeated from the heart, "Here am I, Lord; send me," joining then in the consecration hymn, "I'll go where you want me to go, dear Lord."

There were a large number of brief prayers of consecration, many praying at once, followed by many brief testimonies grouped around the opening sentence of our pledge, our watchword for the coming two years. One returned aviator spoke of the way a beginner is trained, merely keeping his hands on the wheel while his trainer sitting behind him guides the air-ship through the air; and of an aviator who lost his life because under such conditions he insisted upon moving the wheel himself. Every one of us makes just such a fatal mistake when he tries to guide his own life.

Another speaker noticed a line quoted by an Endeavorer, "Thou wilt guide me with thine eye," and noticed at the same time how Dr. Clark, standing in front of the audience, guided the Bible verses and testimonies, indicating by a glance who was to speak next. Thus our Father gives us the clews of our lives, tells us when to speak, when to act, when to keep quiet.

Another returned soldier spoke of his experience in losing his sense of individuality in the army; he was swallowed up in the service of his country, just as every true Christian loses himself in the service of God.

CHAPTER XI.

BANQUETS AND BANQUETS.

The Experts' Banquet.

THE fine Y. M. C. A. building of Buffalo furnished a comfortable room for the banquet of the Christian Endeavor Experts. Every seat was filled with a crowd of some 250 Endeavorers who have taken the Expert training, passed the examinations, and are privileged to add to their names the letters, "C. E. E."

All through the bountiful meal the room rang with merry Christian Endeavor cheers, especially marked being the capital Christian Endeavor versions of the popular war tunes, and the cheers of the cities that want the Christian Endeavor Conventions of 1921 and 1923.

The toastmistress of the evening was Miss Margaret M. Bradt, the bright field-secretary of the Minnesota Christian Endeavor union. She told the story of the Little Red Hen who found the grain of corn, and applied it ingeniously to the various speakers and their themes.

First she called upon Amos R. Wells, who gave the Experts an impromptu lesson in the spelling of "Expert," the Endeavorers suggesting the significance of the letters. Thus "Ex" stands for *Experience;* no one can become expert from a textbook merely. P means *Prayer,* for no one can be really expert in Christian work without God's help. E stands for *Enthusiasm,* the zeal which holds us true to the purpose which calls for expertness. R signifies *Repetition;* we become expert by doing things over and over. T means *Teachableness,* for the true Expert gets help and hints from all other workers.

President Wood of the Mississippi union, forcibly described the young folks who do not want expertness, who are willing to live weak and inferior lives, to use false excuses for not doing their duty, who can find plenty of time for the movies but no time for the church.

President Stauffacher of the Iowa union, said that Experts occupy a strategic position in Christian Endeavor progress. Our society has made so much progress in past years without Expert training; how much more may we expect from Christian Endeavor with this new expertness? We are living in the day of experts. The subdivision of labor and opportunity for training has made the expert in industry possible. So in the church, if we want true expertness, we must subdivide the work and insist upon a training in expertness.

Dr. Robert Hume, ex-president of the India Christian Endeavor union, told of some of the curious committees among Indian Endeavorers, especially the committee on nail-cutting and the committee on picking up paper and rags about the church

Dr. Pettee, president of the Japan Christian Endeavor union, told of the common Japanese Christian Endeavor committee on lectures, which looks after the educational side of the work, and especially after the lectures of which the Japanese Endeavorers are so fond. The Japanese are hero-worshippers, and expect leaders to do all the work. They need to learn how to lay responsibility on a large number of people, and Christian Endeavor is doing much to teach them this.

This banquet was one of the most delightful and profitable of the Conference gatherings.

For the Juniors.

It was a merry company and a large one that filled the tables at the Junior workers' supper on Wednesday evening at the Buffalo Consistory. Here Miss Grace F. Hooper, national Junior superintendent, was supreme and performed the duties of toastmistress with her usual enthusiasm, tact and charm.

The toasts were based on the parts of an automobile. Mrs. William F. Martin, Junior superintendent of the Illinois union, opened with an enthusiastic talk on "The Brake." The brake is the pledge. If we can impress the ideas of the pledge on the hearts of the boys and girls we shall greatly help them to avoid life's dangers.

"The Spark Plug," was the subject to which Miss Sylvia L. Tschantz, Junior superintendent of the Ohio union, addressed herself. Without the spark plug, the machine will not go; the car is dead. The speaker strikingly drew the parallel between the spark plug and *prayer,* which brings us the power of God. She told of superintendents who divide the Juniors into six companies and pray for a different company every day and for all the companies on Sunday. In Ohio they have a superintendent's prayer-list according to which the superintendents pray for each other by name—so many names each day.

Miss Ethel Brown of Michigan, was given "The Lights" for her subject. She brought out the thought of shining for Christ, each according to His power, and pleaded especially for missionary education among the Juniors, that their light may shine in far-off lands. They can shine through giving their money and through making articles that may be sent to the mission fields.

"Gasoline" was the topic treated by Miss Grace Cole of Rhode Island. The gasoline is the thing that makes the car go. We must give the Juniors things to do if we are to hold their interest, big tasks for all. The speaker told of the Efficiency Chart, which suggests work enough for everybody.

Miss Agnes E. Baker, Junior superintendent of the New York union, discoursed in poetry on "The Crank," giving a clever description of the various things for which cranks are good.

The next speaker was Miss Emma McClaran, president of the Oklahoma union, who formerly was Junior superintendent, and who spoke on "Wheels." As the automobile cannot go without its wheels, neither can a society make progress without committee organization and work. Through such work the Juniors receive a splendid training that fits them for larger tasks. Juniors must learn, by doing it, to conduct a business meeting and other kinds of work. Get the right boys and girls in the right place, so that the society automobile does not have different sizes of wheels on it.

Mrs. Francis E. Clark, booked to speak on "The Clutch," received an ovation. She confessed a predilection for her own feet rather than an automobile as a means of locomotion. Feet never run away or scare their passengers. Witty remarks aside, and they were happily inspired. Mrs. Clark found in "Faith" a beautiful parallel to the clutch. It is by faith that we get a grip on God; and this is the kind of clutch that we need. We must have have three things in Junior Christian Endeavor. First, faith in God. Second, faith in ourselves. We must believe that we can do whatever it is our duty to do. And third, we must have faith in the boys and girls, who, if they are not saints, are nevertheless *called* to be saints, and who may yet be saints if we have patience with them.

Miss Hooper called upon Lieutenant James Wray of the aviation corps, who brought a helpful message and urged, in the words of a friend, that we let all our blow-outs be socials to save. Feelingly he told of the suffering of the boys and girls of France under the heel of German tyranny. The children of the whole world need us.

The Intermediate Banquet.

The Intermediate banquet on Friday evening was a hilarious affair. It started off right, with Paul C. Brown, national Intermediate superintendent, presiding. It was a real Intermediate affair full of pep and the spirit of happiness. Paul C. Brown demonstrated his ability as choirmaster, toastmaster, conference leader, and Intermediate Expert; in fact, as master of the situation. He magnified his office. For him Intermediate Christian Endeavor is the biggest and best thing in the whole convention.

Miss Agnes Baker, in welcoming the young people, said that in Erie County they believe in I. C. E., whether it is cream or otherwise. In this case it was *cream*.

Mr. Brown stated that there are abount 50,000 Intermediate Endeavorers in the United States, all of them of high-school age.

The first talk was by President Rusk of the Georgia union. He spoke to the Intermediate leaders, and urged them to make paths along which young people may travel.

W. Roy Breg, Southwestern field-secretary, referred to the fact that there are more Intermediate societies in California than there are Junior societies. At the last California convention more than 1,000 Intermediates registered. Texas has taken up Intermediate work with all her heart, and in the past few years has doubled and doubled and doubled yet again the number of Intermediate societies. Mr. Breg told of the great interest in religious matters which California Intermediates have, and he pointed out the opportunity we have through the Intermediate society to fashion the religious ideals of youth.

Miss Jessie Grider Butts, Texas's field-secretary, formerly an Intermediate superintendent, suggested, as she saw it, the biggest problem in Intermediate work. It is that of finding the right kind of superintendent. There are three essentials for a superintendent: grace, grit, and gumption. The superintendent must be consecrated, a Christian all the way through. He must also have grit. He must not be weak and wabbly, but must stand erect for his ideals. He must have plain horse sense, gumption. He must know the Intermediates and be ready to give wise counsel when wanted. Advice is always cheap when it is not asked for.

Paul Brown knows how to make an appeal and draw in the net. It was encouraging to note that more than a score raised their hands to indicate that they are willing to become Intermediates, if God opens the way.

Mr. Haines A. Reichel, formerly of Ohio, now field-secretary for Pennsylvania, offered a personal testimony. He said that at the time when he was called to work for the Intermediate societies of Ohio he did not believe in Intermediate work. But faith soon came when he got into touch with the Intermediates themselves. The work became absorbing, easy, a joy. The best way to become a good Intermediate superintendent is just to become a superintendent and try to do your best.

Rev. J. Vincent Knight, president of the North Carolina union, made a rousing talk on the responsibility of Intermediate work. There is no more important period in young people's lives than the 'teen age. They are being led, and they are learning to be leaders. They are the making of the church of to-morrow. A pastor of a Southern church said that in a period of years 329 young people had given themselves to life-work for Christ, and all of them except two or three were Endeavorers, a large number of them in the 'teen age when they made the great decision. The Intermediate age is the fluid age, the age of decisions, and therefore the age of greatest opportunity.

Lieutenant James Wray was the last speaker. It was an enthusiastic word he spoke that linked up the Junior and Intermediate societies. The Intermediate superintendent's work is no easy job, but it is a great work all the same.

The banquet demonstrated great interest in Intermediate work, and gave promise of the organization of not a few societies when the young people who caught the vision at this meeting go back home to make their dreams come true.

The Tenth Legion Banquet.

Again the Y. M. C. A. hall was crowded on Friday evening for the Tenth Legion banquet, over which Mr. Walker presided with his invariable good cheer and good sense.

President McClusky of the Massachusetts union, the first speaker, spoke of the joy of giving. "No matter how much the giving makes my pocketbook feel as if an elephant had stepped on it," he said. "I never have regretted a gift, or felt that I was a loser or a sufferer from it. The giving of money is the smallest part of our giving. How about a tenth of your time? Do you give that freely and gladly for the Lord's work? We are to give our own selves to the Lord as well as our money. Tithe your time."

President Hewetson of the Ontario union, spoke of the need of tithing to support a greatly needed field-secretary in Ontario, of his own enthusiasm over the practice of tithing, and his hope that he may persuade his business firm to tithe.

President Miley of the Virginia union, spoke of the financial gains of tithing. Many are afraid that tithing will relieve them of their money, but really it will multiply their money, in most cases, by two. Mr. Miley gave some striking examples of this truth. Tithing pays from every point.

Dr. N. B. Grubb, Mennonite trustee of the United Society, told of a young man who set out to prove whether tithing paid or not. He soon began to branch out in business, became more and more prosperous, ascribed it all to his practice of tithing, and now is not only tithing his income but tithing his business. And his business amounts to two million dollars a year.

Another trustee brought his message, Rev. T. M. Funk, representing the Church of God. Out of an experience of years he believes in tithing, because it is systematic, affords him ready money for God's work, is not burdensome, and brings in more money with God's blessing. It is God's plan for church financing. If all Christians should practice tithing, we should solve all our church financial problems without church socials and entertainments.

President Newsom of the Tennessee union, made a forcible plea for the wise use of money as against making it an idol. If we use money rightly, we bring real joy to the heart of our Saviour, to other men, and to our own hearts.

Field-Secretary Farrill of Wisconsin, insisted that tithing is sound economic principle. It is not enough to be a proportionate giver, or a systematic giver. He once knew a systematic giver who gave $1.25 a year to the church regularly. What we need is liberality, plus system.

Dr. Hume said that tithing has begun in India, and tithing of time as well as of money. Dr. Kendall, trustee from the Christian Church, made the point that tithing is the basis, not of *giving*, but of *paying*. We are simply paying our debts. With a very straightforward and impressive personal testimony to the blessedness of tithing by Mr. Walker, and the closing prayer by Amos R. Wells, this very practical and cheering supper came to a fine conclusion.

CHAPTER XII.

INTERCHURCH WORLD MOVEMENT.

ELMWOOD MUSIC HALL.

Friday Evening, August 8.

"IN the twentieth century we are coming into the gospel of love," said Dr. Rufus W. Miller in opening the devotional service in Music Hall on Friday evening, and from that thought sprang the Bible-reading and prayer.

After the singing of a beautiful solo, Dr. Clark presented a banner, on the part of the Dixie Endeavorers, to the Georgia union, as a reward for making the best record of all Southern States during the past year.

Dr. Clark then pictured in moving terms the splendid Christian Endeavor career of General Secretary Shaw, which has for so many years moved parallel to our society's history. He spoke of his resignation presented at the trustee meeting of the afternoon, the resignation accepted by the trustees most sadly and regretfully, and because he insisted upon it. Ever since his terrible automobile accident of three years ago Secretary Shaw has been heavily handicapped for his work, and has kept it up bravely and faithfully, but under severe difficulties.

Secretary Shaw's resignation does not take effect until the first of next year, by which time, it is hoped, a successor can be found. And, be it remembered, Dr. Shaw has not resigned from Christian Endeavor by any means. He will still be the treasurer of the World's Christian Endeavor Union. He will still be high and active in the councils of Christian Endeavor. And as the publisher of *The Christian Endeavor World* he will contiuue to do a work for our society of far-reaching, endless import.

Turning to the special theme of the evening, we thought of "The Interchurch World Movement as a Minister Sees It," the speaker being that pastor of far-seeing vision, Rev. Ernest H. Tippett, D.D., of Cleveland. He spoke from the viewpoint of one who has just returned from the horrors of the front in Europe. On his return he had found one Christian Endeavor desire and prophecy accomplished in the abolition of the liquor traffic, and another, the abolition of sectarianism, on its way to accomplishment through the Interchurch World Movement.

Dr. Tippett gave a stirring account of the American soldier as he saw him—millions of him—in France. Now that these noble young men have come back, or soon will come back, it is for the church of God to set before him a programme that will still his blood and call out the best in him, even as it has been called forth by the summons of Justice and Liberty in the Great War.

No one knows more about "The Interchurch World Movement, What It Is and How It Works" than Dr. J. Campbell White, the next speaker. "If twenty-three nations could coöperate to win the war, cannot all the denominations coöperate to win the world?" he asked. The Interchurch World Movement is a serious effort to map out what remains to be done to complete world evangelism, and then to do it. It is working through the existing denominational agencies, and through interdenominational organizations like Christian Endeavor. It originated in the Southern Presbyterian mission boards, who brought together a great company of mission boards, and all of these, meeting in New York City, approved the plan unanimously. The plan is instituted for five years, but Dr. White is sure that the churches will never go back to the old separate action. He believes that this is the most significant and important movement since Christianity began; and who will not agree with this judgment?

The Interchurch World Movement is not working for church union; let that come as it will in its own good time. This great new movement has regard to the deep spiritual union which underlies all separate denominations.

Dr. White has long prayed for the coming of the time when the gospel shall be brought to all mankind. His hope of that time is immeasurably greater to-day than ever before. But he declares that to carry out an adequate programme we are going to need a hundred thousand new recruits, with the most thorough training for them all. "The power of the church is simply measureless if we allow Christ to project His powers through it."

"Four million American soldiers. That shows what we can do if we try. Four thousand male missionaries, only one for every thousand soldiers. Three thousand Endeavorers in this Conference. Only one foreign missionary for each company of Christians of that size. This shows what we are doing, now that we are not trying." The Interchurch scientific survey will show what we must do. And then we will try it, and do it.

Mr. S. R. Vinton gave great pleasure and profound instruction to the Conference with his superb lecture, "A World Survey in Picture and Story." This lecture was a great feature of the recent celebration of the Methodist Episcopal Centenary. It consisted of a very large number of splendid pictures illustrating missions all over the world. There were many striking diagrams and charts showing the vastness of the missionary need, and the comparative insignificance of present accomplishment. Many impressive sentences, by some of the greatest of living thinkers, were thrown upon the screen, all urging the necessity of missionary progress. And there were scores of beautifully colored pictures, representing all sorts of mission scenes, and all were full of meaning. No one could be present during that hour and not feel his missionary conviction strengthened and deepened.

CHAPTER XIII.

THE DENOMINATIONAL RALLIES.

United Brethren Rally.

THE United Brethren rally, led by Dr. O. T. Deever, secretary for Christian Endeavor in his denomination, discussed the best things each had got from the Buffalo Conference. Many of those present resolved to go home to promote Interchurch and Junior work, to organize Alumni fellowships, and to carry the message of the Buffalo meeting to their own and other societies. More time in State conventions for denominational rallies was also recommended.

Dr. H. F. Shupe, editor of *The Watchword*; State Secretaries Spafford of Michigan, and Hetzler of West Virginia; Dr. J. S. Kendall, superintendent of the Tenth Legion of Ohio; Rev. Mr. Strayer and Rev. Mr. Fulton, leaders in United Brethren Christian Endeavor work, and union and local society workers presented valuable suggestions.

Christian Church Rally.

It was an earnest, enthusiastic group which gathered at the rally of the Christian Church delegates.

Plans for deepening and developing the spiritual and devotional life of the societies were discussed, and the new goals of the church Christian Endeavor department were discussed and accepted.

It was thought wise, in accepting the goals set before the Conference by Dr. Clark, to link these goals with the aims and goals of their own Forward Movement.

A discussion with regard to Christian Endeavor in the rural churches brought out some helpful thoughts and plans. The aim to organize a Christian Endeavor society in every rural church where one does not already exist was expressed.

It was also decided to urge the issuing of challenges to increase their Efficiency contests between conferences.

Considerable discussion, all favorable, was engaged in with regard to putting a field-secretary of Christian Endeavor into the field for full-time service.

Among those present were Rev. F. G. Coffin, D.D., president of the New York Christian Endeavor union; Rev. J. V. Knight,

Dr. John Timothy Stone.

George E. Gowing,
Director of Music.

Duncan B. Curry,
Chairman of the All-South Extension Committee of Christian Endeavor.

S. J. Theodore,
Madura, India.

president of the North Carolina State union; Rev. Mr. Loftus, Christian Endeavor secretary of the Rhode Island and Massachusetts Conference of the Christian Church.

The Presbyterian (U. S.) Rally.

Southern Presbyterian Christian Endeavorers at the Buffalo Conference had an enthusiastic and successful denominational rally Saturday morning at the hour designated on the programme.

Quite a number of delegates had come from various Southern States, and a goodly number of these were from Presbyterian societies.

The general superintendent of Young People's work, Rev. Gilbert Glass, D.D., conducted the rally, and outlined the policy and programme contemplated for the coming year.

Three salient features of this denominational policy were emphasized as follows:

1. The new Southern Presbyterian Christian Endeavor Standard, prepared by the Young People's Department at Richmond, and indorsed by the General Assembly at New Orleans in May for Young People's and Intermediate societies.

A wall-chart of this standard is being prepared and will be sent to all societies early in the autumn, together with explanatory leaflets.

2. The Young People's Council for the local church, also adopted by the General Assembly, as an agency of supervision and correlation of the work of different young people's organizations in the church. Literature regarding this newly projected plan was promised in the near future.

3. The new progressive programme of the Southern Presbyterian Church, and the part to be played in this large movement by Christian Endeavor and other young people's societies.

The Young People's societies have been asked to contribute $100,000 annually toward the three-year budget of $12,000,000 for missions and church benevolences proposed as part of the progressive programme.

There was a spirited discussion of these goals, and the programme was unanimously indorsed by the group. Plans for carrying on a vigorous campaign for promoting the programme were discussed, and the hearty coöperation of All-South, State, and district Christian Endeavor leaders was assured.

Southern Presbyterian Endeavorers were well represented at the Buffalo Conference, which was a great and, in some respects, an epoch-making gathering.

The Presbyterian (U. S. A.) Rally.

At the time appointed for denominational rallies about 175 Presbyterian young people came together to consider their obligations as Presbyterians, and ways in which these would fit into their Christian Endeavor work. The leader of the rally was Rev. J. S. Armentrout, director of teacher-training and organized Bible-class work.

Two special matters were presented by the leader for consideration, question, and discussion.

The Presbyterian young people's programme was discussed first. Item by item it was shown to call for the best efforts of the young people and to lead to large results in furthering the things for which Christian Endeavor stands.

This programme is presented jointly by *Young People's Work* and the woman's missionary boards of the Presbyterian Church, and so is the denomination's challenge to the loyalty of her young people.

The second matter presented had to do with the young people in the New Era Movement. The large place that they must have in this movement was suggested, and some plans were suggested and developed. That the enlarged demands for service being made through this organization are being met and answered cheerfully was abundantly shown by the enthusiastic interest of this group of young people.

Seventh Day Baptist Rally.

The Seventh Day Baptist rally was held in the adult Bible-class room of Delaware Avenue Baptist Church, Rev. William M. Simpson presiding. Mr. Simpson presented the proposed programme of the denominational Young People's board, and spoke of the inter-relation of denominational and interdenominational standards. Rev. William L. Burdick, president of the denomination's General Conference, told of the plans for young people's work at the annual session of that conference, to be held at Battle Creek, Mich., August 19-24, 1919. The financial budget was discussed, and plans were made for doing more to extend Chri tian Endeavor.

United Evangelical Rally.

The representatives of the Keystone League of Christian Endeavor met in the Covenant Presbyterian Church, with Rev. W. E. Peffley, general secretary of Young People's work, in charge. The rally was distinguished for the representative Christian Endeavor leaders present, as Dr. D. A. Poling, associate president of the United Society; Rev. S. B. Vandersall, general secretary of the Ohio union, and Rev. C. H. Stauffacher, president of the Iowa union.

Rev. S. B. Vandersall, in a five-minute speech, laid stress upon the denomination's opportunity and obligation to enter the large world-movements of to-day. "The Keystone League of Christian Endeavor must be in touch with the outside world, especially with the United Society of Christian Endeavor. Coöperation with the big things is our opportunity and our obligation."

Rev. C. H. Stauffacher delivered an enthusiastic address on the importance of leadership in denominational as well as in general Christian Endeavor work. The report of the work of the denomination in Iowa shows signs of general improvement.

Rev. P. E. Smoke of the Ohio conference, presented the plans of the "Forward Campaign" of the denomination, and showed how the Keystone League of Christian Endeavor is especially adapted to meet the conditions of this campaign. The campaign includes Bible study, Quiet Hour, stewardship, and church attendance.

Dr. D. A. Poling presented a brief outline of the new standards of Christian Endeavor, and showed how well the United Evangelical Church ideals will fit into the programme of Loyalty. Dr. Poling, having been reared at the family altar in a minister's home, told how he solved the problem of suitable time for such an altar in his own busy household. The only time when the family group could be together was at mealtime; and that was the time used for the family altar.

The rally was characterized by the unanimous decision to work for the attainment of the new goals during the next two years.

Friends' Rally.

A conference for Friends' young people in attendance at the Buffalo Christian Endeavor Conference was held in the Sunday school room of the Richmond Avenue Church of Christ. Rev. O. W. Carrell, vice-chairman of the Board of Young Friends' Activities, and one of the trustees of the United Society, served as leader of the conference.

The general topic was that of the relation of Friends to the Christian Endeavor movement. It was the feeling that Christian Endeavor can be largely adapted to their needs as a denomination and ought to be so adapted as most largely to minister to the needs of their young people and churches. Friends believe that more serious care should be taken by their Christian Endeavor societies, so that such results may be realized.

It was the feeling that there is much in the Christian Endeavor movement as a great interdenominational movement that Friends need.

They particularly feel that Christian Endeavor is needed at this time when the Interchurch World Movement and various coöperative movements among the denominations are attracting attention.

It was also their earnest prayer and desire to make their contribution as Friends to the life of the Christian Endeavor movement and to the larger life of the denominations as a whole.

They hope to adapt the plans of Christian Endeavor more fully to their condition and needs as a denomination—to emphasize and apply to daily life and to social relationships the fundamental principles of Christian Endeavor, especially that principle, "Whatsoever He would like to have me do"; and to enter heartily upon the Loyalty Campaign for a new loyalty to Jesus Christ, His principles, and the church.

A. M. E. and A. M. E. Zion Rally.

The denominational rally of the A. M. E. and A. M. E. Zion, C. M. E. and Baptist, was held Saturday morning at the Michigan Avenue Baptist Church, Professor Aaron Brown, secretary of the Christian Endeavor Department of the A. M. E. Zion Church, presiding.

Approval was given to Dr. Clark's suggestions of a two-year campaign for a net gain of fifty per cent in the membership of every local society, for organizing a Christian Endeavor society in every available church, a definite goal for denominational missionary gifts, a field-secretary for at least part time in every State, and a sustaining Alumni Association in every union.

A unanimous vote was taken favoring coöperative union of the denominations represented in the meeting. Three-minute speeches were made by Dr. E. H. Hunter, Norfolk, Va.; Mrs. F. E. Hebbons, New York City; Dr. C. B. Lawyer, Cambridge, Mass.; Professor J. W. Eichelberger, Warren, Ark.; Dr. J. W. Brown, New York City, and Mrs. N. E. Taylor, Tampa, Fla.

Among those present taking part in the discussions were Rev. N. H. Bexley, Mr. Lloyd Mitchell, Mrs. A. G. Banks, Mrs. B. B. Gilmore, Miss Mary B. Mason, Mr. James R. Ross, Mr. F. W. Spriggs, Miss E. C. Kinch, Miss E. Frances, Mr. J. R. Moss, Miss Grace Mosley, Mrs. H. B. Eichelberger, Rev. J. W. Gray, Miss Leona Mitchell, and Miss L. M. Rashming.

The Baptist Rally.

To ensure the presence of Dr. Erb, one of the foremost Young People's leaders in the church, the Baptist rally was held on Wednesday instead of Saturday afternoon, the time of the other denominational rallies.

About seventy-five Baptist delegates gathered in the Delaware Avenue Baptist Church and were inspired for an hour by the eloquence of Dr. Frank Otis Erb, editor of Baptist Young People's publications, Philadelphia. The daring programme of American Baptists was outlined, namely, $100,000,000 from Northern Baptists, $75,000,00 from Southern Baptists, and $25,000,000 from our colored brethren, a total of $200,000,000 within the next five years. Great emphasis was also put on the personal service or, as Endeavorers call this enlistment, "Life-Work Recruits." Dr. Erb quoted Wu Ting Fang's recent message to us, "Send missionaries to China!"

Dr. Hume of Ahmadnagar, India, gave a rousing address on conditions among the 317 millions of India, particularly interesting his Baptist audience in the wonderful work done in the Telugu and Assam missions. Dr. Hume said it was no longer necessary to preach in Indian mission stations about "One God." The people assume it.

Mr. H. N. Lathrop, of Boston, clerk of the United Society of Christian Endeavor, spoke from a layman's standpoint on the $100,000,000 campaign. He spoke of the certainty of "going over the top" with this because the job was big enough to awaken every Baptist layman in America to his immediate duty and opportunity. Mr. Lathrop approved the challenge for personal service, and pleaded for a tithed time as well as a tithed pocketbook.

Baptist Endeavorers now have the spirit, vision, and enthusiasm to adopt this great programme of the denominational leaders, and they say to each other and to the world, as Dr. Clark said in his key-note address, "Come on!"

The Congregational Rally.

The Congregational rally, which was held in the North Presbyterian Church, attracted a large company of young people. Rev. Frank M. Sheldon, D.D., secretary of the Congregational Education Society, who led the meeting, gave an exhaustive talk on the opportunities for service that the church presents to its young people. He described the numberless challenges that come to young people to go to the mission field, home or foreign, with the gospel message, to serve the Master as medical missionaries, or nurses, or in social service, and so on. Then he told of the opportunities for education for these lines of service, opportunities for special training for all sorts of Christian activity. The church needs consecrated men and women to-day. It opens up great fields of service that are inspiring young men and women everywhere. Dr. Sheldon gave many illustrations of the points he raised. He spoke very sympathetically of Christian Endeavor, what it has done, is doing, and may yet do.

Dr. James H. Pettee, D.D., who for forty years has been a missionary to Japan, gave a glowing account of the work of the Kumiai church, which is practically the Congregational Church in Japan. This great body is controlled entirely by Japanese, and in it Christian Endeavor is particularly strong. All the Japanese who have been presidents of the Japan union have belonged to this denomination.

In the conference following Dr. Sheldon's talk the question of dancing was introduced by a pastor who said that his young people wanted the privilege of dancing in one of the church halls. The discussion brought out the fact that the dance question is a vital one in many churches. In some cases the young people it was said, declared that they wanted to dance, that they preferred to dance in the church, but if the church refused them the opportunity they would dance anyhow, somewhere else.

Dr. Sheldon gave a brief history of the dance, and spoke very emphatically against some of the modern dances which are vulgar and unchristian.

Dr. William Shaw said that if churches really opened their halls to the social dance, it should be placed in charge of the deacons, who could act as chaperons, and the responsibility should not be loaded upon the Christian Endeavor society.

The Methodist Protestant Rally.

The Methodist Protestant rally was led by Rev. A. G. Dixon, D.D., general secretary of the Board of Young People of the Methodist Protestant Church. He emphasized: 1. The Quiet Hour. 2. The Tenth Legion. 3. Life-Work Recruits. 4. Coöperation. 5. The Forward Movement Campaign outlined by Dr. F. E. Clark in the opening address of the Conference.

Rev. J. F. Cowan, D.D., and Mr. Bassford, both of whom have been engaged in Y. M. C. A. war-service, added much to the rally by their timely remarks. Miss Ella May Wilson, a former missionary to Japan, was also present.

Mr. Carroll M. Wright, president of the Maryland Christian Endeavor union; Rev. L. W. Gerringer, president of the North Carolina Conference union of the Methodist Protestant Church, and Rev. L. C. Little, field-secretary of the All-South Christian Endeavor Extension Committee, were all present and added to the enthusiasm and inspiration of the rally by their timely remarks.

The Methodist Protestant Church may be counted on to do its full share to help put over the programme set forth in the Buffalo Conference.

Reformed Church in the United States Rally.

The delegates and friends representing the Reformed Church in the United States and the Reformed Church in America met in rally in the parlor of the North Presbyterian Church. The rally was under the direction and leadership of the Rev. G. H. Gebhardt, who made a splendid leader.

The programme was purely denominational and consisted of a number of very interesting short addresses, together with the greetings from the several States represented.

Rev. Rufus Miller, D.D., spoke on the Interchurch World Movement, and especially the Forward Movement of the Reformed Church. Mr. Clarence C. Hamilton of *The Christian Endeavor World,* spoke on Christian Endeavor Summer Schools, and their relationship to the church. Mr. A. J. Shartle of the United Society of Christian Endeavor, spoke on the Young People's work in general and dwelt especially on the Young People's Department in the Reformed Church. There were many splendid testimonials given as to the value of Christian Endeavor in the Reformed Churches of to-day.

The meeting was very enthusiastic and the splendid reports given at the rally, together with the inspiration caught from the addresses and fellowship will not only long be remembered, but are also very helpful in the work of Young People's societies in the Reformed Church.

CHAPTER XIV.

A RECREATION RALLY.

NIAGARA FALLS, CANADA.

Saturday Afternoon and Evening, August 9.

ABOUT a thousand Endeavorers packed themselves into the electric cars and took the long ride from Buffalo to Niagara Falls. They were a jolly set of travelers, and many were the jokes perpetrated on the way.

Arrived at the Falls, we scattered all over the reservation, and saw the world's greatest natural wonder from all angles in the thoroughgoing Christian Endeavor way. Then we made our long way over the bridge to the lovely Victoria Park on the Canadian side, and held an outdoor rally around the speakers' stand.

We sang the International Hymn, all standing, and President Stauffacher of Iowa, led us in prayer. Then Dr. Clark, who presided, let us know that he can never be President of the United States because he was born on Canadian soil,—a statement heartily applauded by our Canadian friends. Dr. Clark did *not* say, however, that he was once nominated for President by one of the reform parties!

Dr. Clark spoke of the warm welcome he has received in all parts of the British Empire, all over the world. He spoke most tenderly of his father, who died when his son was only an infant; of his revered mother, a very brave and noble woman; and of his brothers, all of whom, with their parents, lie in a Canadian graveyard.

Rev. W. A. Mactaggart, Presbyterian pastor of Toronto, gave us a most cordial welcome to Canada.

"We are brothers," he declared, "cherishing common aims and hopes. In every great crisis we have stood together. In the Civil War 48,000 young men from Canada responded to Lincoln's call for volunteers, and many of them never returned. The next time you celebrate Memorial Day, take with you, together with the Stars and Stripes, a little Union Jack, that you may place it on the grave of some Canadian boy who died for the ideals that you hold dear.

"Thus also we stood together in the Great War. That Great War is not ended. As Christian Endeavorers we shall still join hands to fight all iniquity and to bring in the Kingdom of our Lord."

Dr. H. F. Shupe, United Brethren trustee, emphasized three Christian Endeavor ideals: salvation, separation, and service,—salvation from sin, separation from all evil, service of all

At the Canadian Rally.

mankind. These were to have been emphasized by Bishop Weekley, who was detained from the Conference.

"We have crossed the boundary line without a jerk. A lady who lived in South Carolina, near the boundary line, was made a citizen of North Carolina by a resurvey, and at once her health began to improve because the climate of North Carolina is considered superior to that of South Carolina! Many of our national distinctions are as fanciful as that notion."

The address was an eloquent plea for a real brotherhood that transcends all boundaries, and unites all true hearts in the bonds of Christian Endeavor.

Dr. Clark spoke of the battle of Lundy's Lane, which took place near where we were meeting. "We are not proud of that battle," he said, "nor of the War of 1812. There has never been another war between the two countries, and there never will be."

Daniel A. Poling then made one of his characteristic talks, bright, winsome, and truly eloquent. "I have never come to Canada," he said, "without finding myself benefited and blessed." He was deeply moved on this first return to Canada since the war, remembering the heroism of Canadian soldiers and the great debt which America, with all the world, owes to them.

"This war," said Dr. Poling, "has in many respects vindicated the Word of God, and certainly it has taught us that victory is 'not by might nor by power,' but by God's Spirit. I am chiefly hopeful to-day that, as we found the spirit that enabled us to conquer in war, so we may find the spirit that will give us the victory in the new days of peace."

Then Poling made a powerful plea for service in patriotism. "To save America," he said, "we must serve the world, and unless we are willing to serve the world we shall never save ourselves. We must be willing to lose much or we shall lose all. I believe in the League of Nations because I believe in my country and her destiny, and because I believe in the Son of God, the Saviour of the world."

All rising, we sang one stanza of "God save the King," one stanza of "America," one stanza of "Blest be the tie," and the closing prayer was offered by Rev. Mr. Hunter, pastor of the African Methodist Espiscopal Church of Portsmouth, Va. Christian Endeavor has never conducted a finer outdoor patriotic rally than this within sight and sound of Niagara Falls.

CHAPTER XV.

THE LEAGUE OF NATIONS.

SUNDAY broke fair and cool, a climax of the beautiful weather we enjoyed throughout the Convention. A large number of the pulpits of this city of magnificent churches were occupied by Conference clergymen. Dr. Landrith spoke at the penitentiary, and congratulated himself and the audience on the smallness of the latter, assuring them that the new régime of prohibition would reduce their numbers almost to the vanishing-point.

The afternoon session at the Music Hall had a great theme, the League of Nations, and a most inspiring audience. The half-hour organ recital by Mr. William Benbow was thoroughly enjoyed, and so was Mr. Gowing's inspiring praise service.

Introducing Dr. Brokaw, Dr. Clark spoke of the Old Orchard Convention, so many years ago, at which Dr. Brokaw was present, when the United Society was formed. Dr. Brokaw, who has been for all these years a trustee of the United Society, then led us in the Bible responses and in prayer.

Most appropriately, the session was introduced by a World Pageant, presented by Buffalo Endeavorers. It illustrated simply but very effectively the work that is being done for the foreign-born elements of our population. Group after group filed upon the stage, came to the front, and sang a familiar Christian song in their own language. A Boy Scout stood at either end of the large platform, and formed the termini of a long line—a tripled and quadrupled line—made up of these Americans who have come to us from across the seas.

First a fine group of a dozen Italian young men from the St. Paul's Methodist Episcopal Church carrying the Italian flag; then, ten men from the Russian Baptist Church; next, a Sunday school class of Polish girls. With hearty applause we received seventeen Chinese young men and boys from the Central Church of Christ and the Y. M. C. A.

Seven young men and women from Hungary, bearing the Hungarian flag, represented the Hungarian Baptist Church. Three little Syrian girls sang a sweet song. Another group of eleven Italian young men came from the first Italian Baptist Sunday school.

A fine treat was given us by a Negro quartette representing the work of the Michigan Avenue Baptist Church. They sang

S. R. VINTON.

REV. R. A. HUME, D.D.

REV. J. B. BAKER, D.D.

DR. J. CAMPBELL WHITE.

DR. JAMES H. PETTEE.

two of the wonderful "spirituals," "He rose from the dead" and "Swing low, sweet chariot," and the audience were eager for still more of their music.

Each of these eight groups sang beautifully, with splendid volume and real skill. It was notable that while the older persons sang in their own languages, all of the children sang in perfectly good American English.

Then, with the stage filled with these friends, a beautiful United States flag and a "church flag" were borne upon the platform, and the last stanza of "America" was sung, followed by "The Star-Spangled Banner."

When the groups returned to their places in the audience, Dr. Clark asked them to rise successively, the Endeavorers giving warm applause to each, and Dr. Clark following with an inspiring word about the growth of Christian Endeavor in that land.

Greetings from India were splendidly presented by Mr. S. J. Theodore, an Indian Christian of the third generation, whose father also was an active Endeavorer. He was glad, he said, to represent the fifth Christian Endeavor country of the world, a land of 2,000 Christian Endeavor societies and 55,000 members, whose activity is indicated by the fact that in the years 1915-17 India added 272 societies and 7,000 members. He spoke of his boyhood's Bible with a Christian Endeavor Quiet Hour pledge pasted on the cover. His mother woke him at six o'clock that he might join the other Endeavorers for prayer in the church at 6:30 A. M.

In an eloquent and strong address Mr. Theodore pictured the services in the Great War, on all fronts, of India's one million volunteers. "India, though not Christian, joined hands with the Christian nations to fight the unchristian principles of a so-called Christian nation." The result has been a powerful awakening of national consciousness in India, with great possibilities of service opening before Indian Christian Endeavor. For one thing, India is launching a campaign to make the empire dry by 1925. How the Endeavorers applauded that!

Rev. James H. Pettee, D.D., for many years the treasurer and for the past two years the president of the Christian Endeavor union of Japan, spoke forcibly of Christian Endeavor in Japan. It has been there for twenty-seven years, and has grown to 235 registered societies (there are really more) with 5,500 members. The Japan union has had for its presidents Dr. Harada (for eleven years head of the Doshishi, a great Christian college); Tamura, godly Christian and trenchant writer; and Osada, the St. John of the non-liturgical pastors of Japan. Dr. Pettee also pictured other prominent Endeavorers of Japan, including the self-sacrificing secretary of the union, Sawaya; including Ishii, founder and lifelong superintendent of the first Protestant orphanage in Japan; also the official head of the strong temperance movement in Japan; also a prison evangelist who has the entrée of eight great prisons and receives constant government grants; also the Japanese head of the Salvation Army—all of these, and hundreds more of conspicuously successful Japanese, are Christian Endeavorers.

On the conclusion of Dr. Pettee's spirited address, which dealt in a large way with some of the most important themes of present-day statesmanship, we all rose and gave three "banzais" for Japan.

Africa, and the great Christian Endeavor work among the African Baptist, Methodist Episcopal, and Methodist Episcopal Zion churches of the United States, were worthily presented by Rev. W. T. Johnson, D.D., African Baptist trustee of the United Society. "If India, Japan, China," he said, "need the Christian message, surely Africa with its teeming millions needs it." Africa needs far more missionaries, and far better support for the workers that are sent. Negroes played their part grandly on the battlefield of the Great War. "I can hear those that returned say: 'We have done our duty. Now bring to us the gospel of Jesus Christ.' "

That eminent Christian Endeavorer, Rev. Samuel Guy Inman, the executive secretary of the Committee on Cöoperation in Latin America, deeply interested the audience in a thoroughgoing and inspiring address on the lands to the south of us.

Latin America consists of twenty countries and requires a continent and a half, yet it is one of the least-known parts of the globe. Mr. Inman prophesied that in ten or twenty years Latin America would be the most-studied part of the world.

The world seeks food, room for overcrowded populations, markets for manufactured goods and surplus capital. Latin America more than any other part of the world can furnish these three things. It has the largest stretch of undeveloped fertile land in the world. All the population of the world could be placed in Latin America and not be a third as dense as the population of Porto Rico. If Argentina were as thickly populated as the State of New York, its eight millions would become two hundred and twenty-five millions. Venezuela has three times the area of Japan, but Japan's population is equal to that of all South America. Cuba has more foreign commerce than China, and Argentina than Japan. Argentina exports more beef and more wool than any other country of the world and stands second in the export of wheat. The whole world is dependent on the nitrate of Chile. The forces of commerce are waking up to these opportunities. New steamship lines are bringing us twice as near Latin America in time. United States banks are being established everywhere. Germany, England and France have for years cultivated Latin America. Japan is heading new steamship lines thither.

The Great War has given Latin America a new openmindedness. It has been accustomed to live by itself. Its religion has turned toward the past, is the foe of education and democracy. It has distrusted the United States, but the world war has ended its prejudice against the land that it used to call "the Great Octopus of the North." The war has shown us willing to sacrifice dollars for ideals. Now the great words in Latin America are "Continental solidarity." In spite of persistent and crafty German propaganda, eight Latin America nations declared war against Germany and seven broke off diplomatic relations with her.

The close of this illuminating address was a full and authoritative discussion of the Mexican question. Mr. Inman declared that a worse than German propaganda is trying to bring our country to intervene in Mexico, thus destroying the most hopeful work done by Carranza and the present revolution, and the largest plans for missionary advance. "The solution of the Mexican problem," Mr. Inman asserted, "is not a revolution to be squelched, but an evolution to be guided."

President Pollock of the European Christian Endeavor union, and Mrs. Clow, president of the British Christian Endeavor union, could not get passage in any steamer to attend this Conference, but Mr. Pollock sent a splendid message, read most appropriately by Dr. Clark in this meeting of wide outlook. That message will be printed later in our paper.

All of these speakers urged far greater missionary work in the lands for which they spoke. The closing address of this rich afternoon was by Secretary Shaw, and brought the meeting to a practical conclusion. After discovering that only eleven persons in the great audience were going as missionaries, Dr. Shaw made a plea that those who could not go should send their money to train native workers through Christian Endeavor. Twenty dollars will train 275 young people for a year in Christian Endeavor. Twenty dollars will pay for one day of the worldwide work of Christian Endeavor. A large number of the Endeavorers rose to signify their intention of aiding this great work by their contributions.

CHAPTER XVI.

JUNIOR AND INTERMEDIATE RALLIES.

FIRST BAPTIST AND NORTH PRESBYTERIAN CHURCHES.

Sunday Afternoon, August 10.

The Junior Rally.

IT was a large and very enthusiastic audience that greeted the boys and girls of the Buffalo Junior Christian Endeavor societies in the First Baptist Church. Spirited songs, applause, and a demonstration by Juniors that know how, were features that older and younger folks will long remember.

Juniors everywhere love Grace F. Hooper, the national Junior superintendent, and the Buffalo Juniors heartily applauded as she opened the meeting.

The animated Juniors, decorated with sashes, ribbons, and emblems, were a picture of anticipation as Miss Hooper presented the most loved and the biggest Junior of them all, Daniel A. Poling.

Dr. Poling in his inimitable way told of the Junior society to which he belonged in his boyhood, and how as a lad swinging his feet while sitting on the front bench in the prayer-meeting room of the church where his father preached he and his companions were taught the first principles of Junior Endeavor. Dr. Poling's description of the little lassie with three braids who played the organ, and the competition of the lads for the attentions of the organist, held both present-day Juniors and those of long ago with admiration and attention.

Junior Endeavor Mother Clark, as the Juniors know her, instantly won the host of Juniors by her description of what Juniors can do. Her vivid portrayal of children's welfare work; of things that interest Juniors, and especially her practical illustrations from door-knobs to microbes, were of intense interest to all. Her closing advice to the Juniors was, "Love God, love one another, train for service in Christ's Kingdom."

A Junior exercise, entitled "The Threefold Opportunity," given by the Buffalo Juniors, was of threefold interest.

The first part, "Opportunities for Junior Superintendents," brought the picture of the indifferent Endeavorer, and that don't-care spirit so often found when a Junior superintendent is needed.

Some of the Juniors at the Junior Rally. First Baptist Church.

Some of the Juniors at the Junior Rally. First Baptist Church.

The final decision of the desired superintendent to take up the work as a labor of love brought the Juniors rushing to the platform with great joy, and with a song, "Win the Boys and Girls for Christ."

Never before was there a better demonstration by Juniors of how a Junior business meeting should be conducted than was given in the second part, "Opportunity for Children's Business Training." Fourteen Juniors were present, with the president and secretary in their places. Strictly parliamentary rules were enforced. Reports of committees were in writing and were illuminating. The description of the Junior Efficiency Chart was a revelation, and the election of new members was a good feature.

"Opportunities for Children's Spiritual Training" was the third feature, and proved very delightful. The Bible quotations by Juniors revealed a thorough knowledge of the lasting things in life, and as Junior after Junior recited their verses, the audience applauded with approval and appreciation.

The closing address was made by Karl Lehmann, Southern-States secretary. Mr. Lehmann gave a helpful blackboard-talk from his book on "Using the Book with the Juniors," and as the boys and girls delved into the mathematical problems suggested by Mr. Lehmann, they discovered some of the great Bible truths.

Buffalo Juniors may well feel proud of their demonstration of how to do things and do them well.

The Intermediate Rally.

One of the interesting meetings on Sunday afternoon, the closing day of the Convention, was the Intermediate rally in the North Presbyterian Church. A large number of Intermediates were present and responded splendidly to the inspiring leadership of Paul C. Brown, Intermediate superintendent of the United Society, who had charge of the meeting.

The devotional service was led by the president of a Buffalo Intermediate society, a young man who splendidly demonstrated the value of Christian Endeavor training. Compared with the dreadful silence of an ordinary church prayer meeting and the length and formality of the ordinary prayer, the sentence-prayer period of this meeting was a delightful contrast. The prayers were the prayers of youth, eager, earnest, simple, sincere, and *brief*. They came thick and fast, with never an instant's pause. They give promise of a rising generation that in years of maturity will know how to pray in public, and pray to the point.

The first message was from Haines A. Reichel, who has been superintendent of Intermediate work in the Ohio union. "Accepting Responsibility" was the subject, Mr. Brown announced, in introducing the speaker and pointed out that one thing that 'teen-age young people must take seriously is just this thing—

responsibility to do "WHATEVER" Jesus Christ would have them do.

Mr. Reichel made a strong plea for the growing boys and girls who cannot sit still in church because it *hurts* them to sit still for any length of time. His personal experiences and illustrations evidently struck fire in the minds of his hearers. They recognized the symptoms. Intermediates are looking for leaders, and cling to them when they find them. Dare we refuse the responsibility to provide leadership for these boys and girls?

The field-secretary of the Texas union, Miss Jessie Grider Butts, drew the thought of her talk from a beautiful book, "The Mind of the Master." Miss Butts made the Master's sincerity her starting-point, and laid upon the hearts of the Intermediates the necessity of being true, faithful, and sincere. There is a high call to-day for genuine Christians whose acts, great and small, ring perfectly true. There are counterfeit Christians in plenty. Intermediates must therefore strive to be genuine, not only in the presence of others, but before the tribunal of their own souls and of God.

Dr. Daniel A. Poling, who was to have been the next speaker, gave up his time to Mr. L. W. Barclay, an earnest Endeavorer, who holds a responsible position—he is director of religious education—in the Boy Scout movement, who had come by special invitation to the Conference to tell of possible ways in which the Boy Scouts of America may help to suggest a supplementary programme of work for Christian Endeavor.

He said that the great problem of to-day does not lie with the high-school boy and girl, but with the eighty per cent of boys who never reach high school, but are swallowed up in industry. *This* is the problem that Endeavorers must face if the children of our cities are to be reached and saved. Tied up with the problem of religious education there is another problem, that of recreation, to meet the play instinct of all children. What must we do to utilize aright the unused time of youth usually spent in play in the city streets, play which often destroys the influence of both home and church and school?

The boy, for instance, has several needs. First he has physical needs. The boy is ninety per cent physical and must be trained through the physical impulses. We must base our programmes on this fact of the preponderance of physical activity. Second, the boy has an emotional nature, the struggle for character through living for ideals. And third, he has a social nature, which seeks friendship.

Scouting, said the speaker, is nothing but an attempt to develop character through physical activity. Its principle is self-activity, giving the boys a chance to do things and make mistakes if necessary. The Scout movement does not attempt religious training, but simply tries to help the church to help the boys by reaching them through their desire for a life of active tasks in the open air.

In this Intermediate rally was struck the new note of recreational activity. No one knows what the result will be. In this rally a seed was planted. What will be the harvest?

At the meeting of State and Provincial Junior superintendents on Sunday afternoon the following goals and monthly service programmes were adopted for 1919-21. It was decided to take as a motto, "This one thing I do," and apply it to the new Junior Efficiency Chart. It was unanimously voted to put on a special drive at once and keep at the drive till every society is hard at work on the chart. Superintendents who had used it were loud in its praise, declaring it the finest thing ever gotten out for Juniors.

CHAPTER XVII.

CHAPTER OF ADDRESSES.

BACK TO GOD.

Address by William Jennings Bryan, read by Dr. Poling in the absence of Mr. Bryan, at Buffalo, N. Y., August 10, 1919, at the International Christian Endeavor Conference.

MR. CHAIRMAN AND FRIENDS:

I am glad to be a guest of the Christian Endeavorers whose good work I have seen in all parts of the world.

Christ with surpassing power of condensation compressed into one sentence the four commandments that relate to man's duty to his heavenly Father, and proclaimed, "Thou shalt love the Lord thy God with all thy heart, and with all thy soul, and with all thy mind."

The paramount need of the world to-day, as it was nineteen hundred years ago, is a whole-hearted, whole-souled, whole-minded faith in the Living God. A hesitating admission that there is a God is not sufficient; man must love with *all* his heart, and with *all* his soul, and with *all* his mind—and to love he must believe. His belief in God must be a conviction that takes hold of his very life, grips every nerve and fibre of his being, and dominates his every impulse and energy. The thought of God is the greatest thought that can enter a human mind or heart, and a sense of responsibility to God the most potent influence that acts upon a life.

Higher education, although indispensable to man, if he is fully to realize the infinite possibilities which God has set before him, brings with it dangers, against which the student should be warned. Chief among these dangers is the development of egotism at the expense of intellect and heart; he may forget that "the fear of the Lord is the beginning of wisdom."

Instructors sometimes encourage scepticism, doubtless without intention, by unduly emphasizing the area of the known and the relative importance of material things, overlooking the fact that "the things which are seen are temporal, but the things which are unseen are eternal."

Some instructors even speak lightly of religion, and by clothing infidelity in the attractive garb of science and philosophy lead their pupils into agnosticism, a tragedy which is the more distressing when we remember that college men not only have a prominence far out of proportion to their numbers, but exert an influence upon a still larger circle whose members look up to them for example.

The Darwinian Theory.

How shall we aid the young man? How can his feet be set upon the solid rock? If he has felt the benumbing influence of the theory that links man in generations with the ape, a theory which has paralyzed religious thought and the usefulness of so many of the intellectuals of the world during the last half century, the poison of this false doctrine must be extracted from his system before he can understand life and his relation to the world. That man bears the image of God and not the likeness

Hon. William Jennings Bryan.

of the animals below him is the foundation-stone upon which he must build.

The theory advanced by Darwin puts God so far away that man loses the consciousness of His presence in daily life. Prayer becomes a mockery if, for millions of years the Lord has never touched a human life or laid His hand upon the history of a nation. If we accept the theory that man has reached his present eminence by the cruel law under which the strong kill off the weak, then, if there is any logic that can bind the human mind, we must turn backward toward the brute if we dare to substitute the law of hate. As Christianity is built upon the doctrine of love and has for its object the establishing of a universal brotherhood, the Darwinian theory can have no claim upon the mind or conscience of a follower of the Nazarene.

And it should not have upon any one else, for there is not one single fact in nature to support it. It is not one "link" only that is missing; all the links are gone. It is as contrary to the law of the universe as it is to the teachings of the Saviour.

Nietzsche's Philosophy.

Have you ever carried the Darwinian theory to its logical conclusion? Nietzsche did, and it made him the most extreme of anti-Christians. I had read extracts from his writings,—enough to acquaint me with his sweeping denial of God and of the Saviour,—but not enough to enable me to discuss his philosophy.

As the war progressed I became more and more impressed with the conviction that the German propaganda rested upon a materialistic foundation. I secured the writings of Nietzsche and found in them a defence, made in advance, of all the cruelties and inhumanities practised by the militarists of Germany. Nietzsche tried to substitute the worship of "the Superman" for the worship of God. He not only rejected the Creator, but he rejected all moral standards. He praised war and eulogized hatred because it led to war; he denounced sympathy and pity as attributes unworthy of man. He believed that the teachings of Christ converted men into degenerates and, logical to the end, he regarded democracy as the refuge of weaklings. He saw in man nothing but the animal and in that animal the highest virtue was "the will to power," a will which should know no let or hindrance, no restraint or limitation.

Nietzsche's philosophy would convert the world into a ferocious conflict between beasts, each beast trampling ruthlessly on anything and everything that stands in his way. In his book entitled "Joyful Wisdom" Nietzsche ascribes to Napoleon the very same dream of power,—Europe under one sovereign, and that sovereign the master of the world,—the very dream that lured the Kaiser into a sea of blood from which he emerged an exile seeking security under a foreign flag. Neitzsche names Darwin as one of the three great men of his century, but, thanks to him for doing so, relieves him of responsibility for the doctrine that bears his name by crediting Hegel with an earlier announcement of the theory. Nietzsche died in an insane ansylum, but his philosophy has wrought the moral ruin of a multitude, if it is not actually responsible for bringing upon the world its greatest war.

Do not cherish the delusion that it makes no difference what a man thinks upon these fundamental things. I met a scholarly man in a Western city a few weeks ago, and, in the course of the conversation, expressed the hope that the treaty to be drawn at Versailles would usher in a permanent peace. "Impossible," he replied, cynically; "men are only animals, and animals will fight." "That," I replied, "is Nietzsche's doctrine"; and he answered with a question, "You surely do not believe in the supernatural?" That is the issue. Is man nothing more than an animal, or was he made in the image of God? Whether the treaty which we await will be the beginning of an enduring peace or merely preparation for another war will depend entirely upon the spirit in which it is written.

If it is animated by the spirit of the Prince of Peace, there is hope; if it is built upon the philosophy of Nietzsche, the future is a starless night.

There is no evidence that there is within any living thing a "pushing" power that can, unaided from above, lift to a higher plane, but there is all about us everywhere conclusive proof of a "drawing" power.

The Drummond View.

Drummond presents this theory. He says that nothing in the mineral world can rise above that world unless something reaches down from above and lifts it up,—and we see evidences of this every spring as the plants raise the mineral up into vegetable life. Drummond points out also that nothing in the vegetable world can rise above it until it is lifted up by a power that comes down from a world above. This, too, we see illustrated when the animals graze upon the grass and lift their food from the vegetable world up into the animal world. By analogy, we reason that that which is in the animal world can be made spiritual only by a power from above that draws men into the higher life, and that this power is a *present* and *living* power, not a power whose activity was confined to ages past.

The Atheist's First Question.

We give the atheist too much latitude; we allow him to ask all the questions and we try to answer them. I know of no reason why the Christian should take upon himself the difficult task of answering all questions and give to the atheist the easy task of asking them. Any one can ask questions, but not every question can be answered. If I am to discuss creation with an atheist, it will be on condition that we ask questions by turns. He may ask the first one if he wishes, but he shall not ask a second one until he answers my first.

What is the first question the atheist asks the Christian? There is but one first question. Where do you begin? I answer, "I begin where the Bible Begins." And where does the Bible begin? "In the beginning God created the heavens and the earth." I begin with God, all-powerful, all-wise, all-loving. I begin with a creative cause that is sufficient for anything that can come thereafter.

Having answered the atheist's first question, it is now my turn, and I ask my first question of the atheist, "Where do you begin?" And then his trouble begins. Did you ever hear an atheist explain creation? He cannot begin with God because he denies the existence of a God. But he must begin somewhere. It is just as necessary that the atheist shall have a beginning point for his philosophy as that the Christian shall.

Where does the atheist begin? I have never known any of them to begin farther back than the nebular hypothesis. And where does that begin? In the beginning? No. It begins by assuming that *two* things existed, which the theory does not try to explain. It assumes that *matter* and *force* existed, but it does not tell us how matter and force came into existence, where they came from, or why they came. The theory begins by saying: Let us suppose that matter and force are here, and then the theory begins to work, and according to the theory, force, working on matter, created a world. I have just as much right to begin with an assumption as the atheist has, and I would rather begin with God, and reason down, than begin with a piece of dirt, and reason up. The difference between the Christian theory and the materialistic theory is that the Christian begins with God, while the materialist begins with dull, inanimate matter. I know of no theory that has ever been suggested as a substitute for the Bible theory that was as rational and as easy to believe.

If the atheist asks me if I can understand God, I answer that it is not necessary that my finite mind shall comprehend the infinite mind before I admit that there is an infinite mind any more than it is necessary that I should understand the sun before I can admit that there is a sun. We must deal with the facts about us, whether we can understand them or not.

The Miracle.

I learned years ago that it is the mystery in the miracle that makes it a stumbling-block in the way of many. If you will analyze the miracle you will find just two questions in it: Can God perform a mircle? and, Would He want to? The first question is easily answered. A God who can make a world can do with it anything He wants to. We cannot deny that God *can* perform a miracle without denying that God is God. But would God *want* to perform a miracle? That is the question that has given the trouble, and it has only troubled those, mark you, who are unwilling to admit that the infinite mind of God may have reasons that the finite mind of man cannot comprehend.

If there is among you any one who has been presumptuous enough to attempt to confine the power and purpose of God by man's puny understanding, let me persuade him to abandon this absurd purpose by the use of an illustration which I once found in a watermelon. I was passing through Columbus, O., some years ago, and stopped to eat in the restaurant in the depot. My attention was called to a slice of watermelon, and I ordered it and ate it. I was so pleased with the melon that I asked the waiter to dry some of the seeds, that I might take them home and plant them in my garden.

That night a thought came into my mind—I would use that watermelon as an illustration. So the next morning when I reached Chicago, I had enough seeds weighed to find out that it would take about five thousand watermelon seeds to weigh a pound, and I estimated that the watermelon weighed about four pounds. Then I applied mathematics to the watermelon. A few weeks before some one, I know not who, had planted a little seed in the ground. Under the influence of sunshine and shower that little watermelon seed had taken off its coat and gone to work; it had gathered from somewhere two hundred thousand times its own weight, and forced that enormous weight through a tiny stem and built a watermelon. On the outside it had a covering of green, within that a rind of white, and within that a core of red, and then it had scattered through the red little seeds, each one capable of doing the same work over again.

What architect drew the plan? Where did that little watermelon seed get its tremendous strength? Where did it find its flavoring-extract and its coloring matter? How did it build a watermelon? Until you can explain a watermelon do not be too sure that you can set limits to the power of the Almighty, or tell just what He would do or how He would do it. The most learned man in the world cannot explain a watermelom, but the most ignorant man can eat watermelon and enjoy it. God has given us the things we need, and He has given us the knowledge necessary to use those things; and the truth that He has revealed to us is infinitely more important for our welfare than it would be to understand the mysteries that He has seen fit to conceal from us.

So with religion; if you ask me whether I can understand everything in the Bible, I answer, "No." I understand some things to-day that I did not understand ten years ago, and, if I live ten years longer, I hope some things will be clear that are now obscure. But there is something more important than understanding everything in the Bible; it is this: If we will only try to live up to the things that we *do* understand, we shall be kept so busy doing good that we shall not have time to worry about the things that we *do not* understand.

The Bible.

When one's belief in God becomes the controlling passion of his life; when he loves God with *all* his heart and with *all* his soul and with *all* his mind he is anxious to learn God's will, and is ready to accept the Bible as the word of God.

I need not present arguments in its support; its claims have been established; the burden of proof is upon those who reject it. Those who regard it as a man-made book should be challenged to put their theory to the test. If man made the Bible, he is, unless he has degenerated, able to make as good a book to-day.

Judged by human standards, man is far better prepared to write a Bible now than he was when our Bible was written. The characters whose words and deeds are recorded in the Bible were members of a single race; they lived among the hills of Palestine in a territory scarcely larger than one of our counties. They did not have printing-presses, and they lacked the learning of the schools; they had no great libraries to consult, no steamboats to carry them around the world and make them acquainted with the various centres of ancient civilization; they had no telegraph wires to bring them the news from the ends of the earth, and no newspapers to spread before them each morning the doings of the day before. Science had not unlocked nature's door and revealed the secrets of rocks below and stars above. From how scantily supplied a storehouse of knowledge they had to draw, compared with the unlimited wealth of information at man's command to-day? And yet these Bible characters grappled with every problem that confronts mankind, from the creation of the world to eternal life beyond the tomb. They have given us a diagram of man's existence from the cradle to the grave, and they have set up sign-posts at every dangerous point along the path.

We turn back to the Bible for the Ten Commandments which form the foundation for our statute law, and for the Sermon on the Mount, which lays down the rules for our spiritual growth. The Bible gives us the story of the birth, the words, the works, the crucifixion, the resurrection, and the ascension of Him whose coming was foretold in the prophecy, whose arrival was announced by the angel voices, singing peace and good will —the story of Him who gave to the world a code of morality superior to anything that the world had known before or has known since, the story of Him who is the growing figure of all time, whom the world is accepting as Saviour and as the perfect example.

Let the atheists and the materialists produce a better Bible than ours if they can. Let them collect the best of their school to be found among the graduates of universities—as many as they please and from every land. Let the members of this selected group travel where they will, consult such libraries as they like, and employ every modern means of swift communication. Let them glean in the fields of geology, botany, astronomy, biology and zoölogy, and then roam at will wherever science has opened a way; let them take advantage of all the progress in art and in literature, in oratory and in history; let them use to the full every instrumentality that is employed in modern civilization; and, when they have exhausted every source, let them embody the results of their best intelligence in a book and offer it to the world as a substitute for this Bible of ours. Have they the confidence that the prophets of Baal had in their god? Will they try? If not, what excuse will they give? Has man fallen from his high estate, so that we cannot rightfully expect as much of him now as nineteen centuries ago? Or does the Bible come to us from a source that is higher than man? Which?

But our case is even stronger. The opponents of the Bible cannot take refuge in the plea that man is retrograding. They loudly proclaim that man has grown and that he is growing still. They boast of a world-wide advance, and their claim is founded upon fact. In all matters except in the science of how to live man has made wonderful progress. The mastery of the mind over the forces of nature seems almost complete, so far do we surpass the ancients in harnessing the water, the wind, and the lightning.

Inventive genius has multiplied the power of a human arm and supplied the masses with comforts of which the rich did not dare to dream a few centuries ago. Science is ferreting out the hidden causes of disease and teaching us how to prolong life. In every line except in the line of character-building the world seems to have been made over, but the marvelous changes by which old things have become new only emphasize the fact that man, too, must be born again, while they show how impotent are material things to touch the soul of man and transform him into a spiritual being. Wherever the moral standard is being lifted up, wherever life is becoming larger in the vision that directs it and richer in its fruitage, the improvement is traceable to the Bible and to the influence of the God and Christ of whom the Bible tells.

To the young man who is building character I present the Bible as a book that is useful always and everywhere. It guides the footsteps of the young; it throws a light upon the pathway during the mature years, and it is the only book that one cares to have beside him when the darkness gathers and he knows that the end is near. Then he finds consolation in the promises of the Book of books, and his lips repeat, even when his words are inaudible, "Yea, though I walk through the valley of the shadow of death, I will fear no evil, for thou art with me, thy rod and thy staff they comfort me," or "I go to prepare a place for you, that where I am, there ye may be also."

What Think Ye of Christ?

And one more word. What think ye of Christ? I do not present Him merely as the highest type of man, but rather as the Bible presents Him, as the Son of God and the Saviour of the world, as He presents Himself when He says, "I am the way, the truth, and the life." Do you have difficulty in believing His divinity? It is because you have measured Him by the rules that apply to man. Take Him out of the man class, and put Him in the God class; and then it will not be difficult to understand Him. Follow the suggestion made by Carnegie Simpson in his "Fact of Christ," and find conclusive proof of His divinity in what He said and did and was.

Measure Him by the task which He came to perform; it was not a man's task. Measure Him by the record He has made. Why, if He was but a man, has not our civilization produced another of His kind? Why are even His enemies compelled to admit the magic of His name and the wonder-working influence of the philosophy He taught? Why are His words as potent to-day as when the fishermen of Galilee became His disciples, as convincing as they were when "the common people heard him gladly"?

Are you in doubt about His power to perform miracles when He walked among men? He is performing them to-day. The Christ who can to-day open the eyes of a young man, who sees nothing but the body and knows nothing but the pleasures that come through the flesh—the Christ who can open the eyes of such an one to the larger vision of the spiritual life could have opened the eyes of the physically blind. Do you question His power to raise the dead? Go into any rescue mission and hear the testimony of those who after years of dissipation and of crime have come under the influence of His grace and been born again; behold the change. The Christ who can take a man from the gutter, one who has fallen so low that even his own flesh and blood have abandoned him, and lift him up, cleanse his heart and fill it with a passion for service—such a Christ could have broken the bonds of the tomb.

I am done. I would call you back to God, to the Bible, and to Christ. If I have succeeded in doing this, I have accomplished my purpose. But I shall be happier still if among you there is one young man or woman whom I have been able to help, one who has been made stronger to resist temptation and whose conception of life's possibilities has been enlarged; for one life, filled with love of God and devoted to the welfare of his fellows, can bring incalculable blessings to a community, a State, a nation, or a world.

CHRISTIAN ENDEAVOR THRIFT CLUBS.

By Dr. J. Stanley Brown, *Secretary, Savings Division,* Washington, D. C.

I am indeed pleased to be afforded the opportunity of presenting for the consideration of this Christian Endeavor Convention the movement for the practice of thrift inaugurated by the Government. To teach the American people the value of saving and the imperative necessity of cutting down waste has made a strong appeal and received unqualified indorsement from men and women in all ranks of business and professional life.

Paul Warburg, a former vice-president of the Federal Board, in a recent speech before the United States Chamber of Commerce states that "the fundamental remedy for all our economic ills lies in thrift." Dr. John Grier Hibben, President of Princeton University, said, "I can conceive of no more valuable contribution which Princeton men can make at the present time to the welfare of our nation than to have a large and active part in creating a general spirit of thrift throughout our country." Samuel Gompers, president of the Federation of Labor, said, "May I say to the 165,000 secretaries of War Savings Societies, 'Yours is a fine work, a necessary work, and one to which every thoughtful person will wish to bring to you help'?"

These opinions are from clear-visioned Americans: one a practical business man; another a man familiar with the success and failure of young Americans; and Samuel Gompers, the voice of American labor.

The Treasury Department feels that, in its endeavor to bring before the people of the United States the value of thrift and to teach them the simple principles of how to save, how to invest, and how to spend, if it can enlist the splendid efficiency of the Young People's Society of Christian Endeavor, this message of the meaning and value of saving will be carried to millions of our American people.

The men who inaugurated this movement believe that thrift has a vital connection with the principles of Christianity; that thrift may be taught with profit and propriety in every church in the United States, through the medium of the sermon, prayer-meeting address, the discussion of thrift problems in the Young People's societies, and in the Sunday school.

It is hardly necessary to state to this Convention the need of the practice of thrift among our people. I have no wish to bore you with figures, but consider this thrift movement for a few minutes from the viewpoint of the wastes common to American life. One billion dollars was substantially the amount of money required before the war to run all the departments of the Government of the United States,—one billion dollars. During the twelve months preceding the armistice the junk salvage of the United States under the supervision of the Government amounted to $1,500,000,000. Enough to run the Government in peace times before the war, and present every man, woman, and child with $5. Think of it, this vast sum was salvaged from the old metal, old rubber, old rags, and waste paper that we usually throw out the back door. It is an actual fact that each truck-load of garbage in New York City contains $50 worth of usable material. In New York City the usable material in the individual garbage-cans ran from 2 per cent. in the poorer districts to 11 per cent. in the wealthy districts. In May, 1918, this statement appeared in one of our magazines from the pen of a physician engaged in the work of the Food Administration. "We all know," he writes, "that we Americans are the most extravagant of all civilized nations, but I own I was amazed at the reckless improvidence of the well-to-do-classes. We counted in 15 garbage-pails in one district no less than five whole loaves of bread. Half and quarter loaves, cooked meats and vegetables, we found in quantities that were almost incredible. Combine that statement with the following and you will realize the significance: If 20,000,000 families waste one slice of bread each day, that waste would be equivalent to a waste of one million loaves of bread each day." Few men and women realize that 85 per cent. of the American people are living from day to day with nothing between them and actual poverty but the daily wages. Eighty per cent. of the men in the United States over sixty years of age are partly or wholly living on friends, relatives, or public charity. It is estimated that 66 out of every 100 die in this country leaving no estate.

These facts, and they are facts, should make the most thoughtless of us willing to listen to words of counsel and advice. The common experience of the race bears witness that the man or woman who is thrifty, who has something to fall back upon in life's emergencies, who has made reasonable provision for old age, accepts the responsibility of his career with confidence, and faces the world in later years with an adequate measure of self-respect. H. W. Hoot said: "I have come to learn that the reason why so many men are in the down-and-out class is the lack of thrift."

In a recent issue of a medical publication, the statement was made that worry is the cause of more breakdowns than any other ill common to the human race. Probably no form of worry is so prevalent among men as financial worry. A man or a woman who is troubled in mind about money difficulties is seldom in a satisfactory mental state to listen with any degree of concentration to the messages of the gospel. On the other hand, the individual who cultivates the habit of thrift secures freedom from financial worries. The very effort he has made to practise thrift has given him a vision of higher and better things that make him more receptive to the messages of the gospel of Jesus Christ. In a similar manner this practice might be applied to a community. A thriftless community, because of wasteful habits, has few possibilities for the growth of religious and civic institutions. On the other hand, a thrifty community has the means to support and advance not only the church, but all civic organizations that labor for the upbuilding of the community's life.

In a discussion of this matter of thrift and saving some weeks ago, a minister, who is deeply interested in the movement, remarked to one of the directors of the Savings Division: "Last night I was thinking over the work of your division. I tried to picture what a thrifty United States would mean to the church. The material prosperity that would come as a natural result of our people being thrifty would remove a great many of the obstacles that hold back the church and other religious organizations from inaugurating a more aggressive policy. "Of course," he continued, "I do not mean by the United States becoming a thrifty nation that all our people would become wealthy. What I mean is that when a man becomes thrifty, when he sees his savings growing, there is removed from his life the worry of how he will support his family if sickness strikes him down, or death, or what will become of him in old age. With a man who has started to save, these dark shadows of the future gradually disappear. He may never become wealthy, but his savings will be sufficient to meet any ordinary emergency. This without question brings a certain peace of mind, and he becomes more easily interested in the things that make for the advancement of the kingdom of God. Almost unconsciously he will become interested in higher ideals of life, he will develop a larger receptivity to spiritual influences, and he will soon appreciate the pleasure of belonging to, and the satisfaction of supporting, those institutions that bless humanity."

That is what this minister believed a thrifty United States would mean to the Christian church. He saw over a hundred million people comparatively free from one of the main worries of life, developing higher ideals of living, becoming more receptive to spiritual influences, and supporting generously the institutions that labor for the upbuilding of mankind.

The whole campaign of thrift springs from a religious motive. Christ taught thrift in the miracle of the loaves and fishes—the remains of the feast must be gathered up that nothing be wasted. In the wise and the foolish Virgins, He taught the folly of individual waste. In the Old Testament, Joseph is shown as an example of thrift storing up the grain in preparation for a future emergency,—a lesson in natural frugality. Of Christ's thirty-eight parables, sixteen relate to thrift and the use of money. In much of His teaching Jesus used as a test the uses to which men placed their wealth, clearly showing the imperative necessity of a correct appreciation of its use.

Bishop Fowler in a sermon said "I believe, on the day of judgment, more people will be condemned for the way they used their money than for any other thing."

The average thrifty man is usually temperate in all his affairs, not only in the use of money, but in time, in health, in everything that pertains to his every-day living. In nearly all cases the thrifty man is a mainstay in the support of the church, and ready to serve in any movement that has for its object the upbuilding of truth and right among his fellow men.

The Treasury Department is confident that if the Young People's societies of the church will get behind this thrift movement, it is only a question of time when the stabilizing effect of thrift will be felt in every department of American life.

As a concrete medium for enabling young people and people of moderate means to save, the Government has issued the Thrift and War-Savings Stamps. These securities bear interest at the rate of 4 per cent. compounded, and the Government of the United States is behind them. Of course, in this matter of thrift, the Treasury Department has no wish to restrict the savings of young people to War-Savings Stamps. Any legitimate investment is satisfactory; the main thing is to save. There is, however, a growing demand on the part of young people and people of moderate means for a security that pays a reasonable interest and is absolutely safe. It is to meet that demand, and make Government securities easily accessible to young and old, that the Government has issued the Thrift and War-Savings Stamps.

In adopting this campaign of thrift it is not the purpose of the Treasury Department to urge our young people to sacrifice, but rather to encourage them to save; not to discourage their spending, but to encourage wise spending; not to restrict the little pleasures of life, but to enlarge our people's independence.

The Treasury Department is endeavoring to encourage the people of the United States—

First, to put aside as their obligation a part of their income for future use.

Second, to invest the money they save in Thrift and War-Savings Stamps—Government securities that are absolutely safe.

Third, to use the rest of their incomes so as to make every penny they spend buy something they really need and want, and which has a full penny's value.

Thus the object of the Treasury Department is to take to every child in the Sunday school, every member of the Christian Endeavor Society, every workman in the factory, every clerk in the store, to every manufacturer, merchant, baker, farmer, housewife, the simple principles of how to save, how to invest, and how to spend.

During the past four years the Christian Endeavor societies have been of inestimable aid to the Government, and to-day the Treasury Department, knowing the willingness of the Christian Endeavor leaders in the past, has no hesitancy in suggesting to this Convention certain methods by which its members can advance this thrift movement. First by introducing the idea of thrift into the subject matter of their regular meetings. This would not mean savings in money alone, but the elimination of waste in health, in education, in useless recreation, and in a score of other ways. The leaders of the meetings, from their practical experience in school or business, would have at hand a dozen illustrations to enforce the value of thrift. They can also inspire by their presence and words their associates outside Christian Endeavor circles.

The Young People's societies will find much interest in a pamphlet published by the Government and especially adapted to their use, entitled "Avenues to Affluence," which discusses such subjects as "Education," "Business Opportunity," "Marriage," "The Future," "Affluence and the Common Good," and several other topics of kindred character. At the end of each chapter are ten subjects for discussion in class. Another feature might be the organizing of a War-Savings Society in each Christian Endeavor society to save money for some special object in connection with their church-work. In one Young People's Society they have organized a War-Savings Society and are saving for a pipe-organ. The investment of the savings in War-Savings Stamps makes it possible for a Young People's society to start a fund that will mature in five or more years to be used for some specific religious purpose.

The Government has the utmost confidence that when the Christian Endeavor workers understand the educational character of this movement, —that there are to be no drives, no stunts, simply a piece of clean-cut educational work,—and the transforming influence the practice will have upon the people of the United States, the thrift movement will not merely have their co-operation, but a definite and active service along the lines I have suggested from every member of the Christian Endeavor.

Next to the great fundamental truths of Christianity, I do not believe there is a more important subject that can be taught to the young people of the United States than this matter of thrift. I feel confident that the officers and members of this Convention will lend their active support to this beneficent and patriotic movement so ardently espoused by our Government.

Y. M. C. A. AMONG FOREIGNERS.

By Paul V. Shaw, *Director Latin America Division.*

When we realize that there are 7,000 foreign students in our country studying in the various colleges and universities, we truly can say that the eyes of the world are upon us. These students are the future leaders of their several countries and constitute one of the most important keys to international friendship. They are the men who will control trade, commerce, politics, and religion and mold opinions. If they receive the correct and true opinion of our people they will be the ones who will interpret it to their own native lands and in that way serve to make stronger the ties of fellowship and friendship between our country and their own. Probably there is no greater opportunity than this to help mold the opinion of the world. The conception which these students form of our country depends entirely on the reception which they are given by those of us who live here. They will not judge America by its tall buildings, by its rapid transit, or by its tremendous strides in science. Their judgment will be based upon their human contacts and the way in which they are treated by the human element. Therefore, the responsibility of giving an accurate and worthy impression of the United States and its ideals rests upon us, and we must take up the challenge which presents itself.

The Christian Endeavor is one of the few Christian influences in cities and towns where there are no colleges and universities or Y. M. C. A.'s or Y. W. C. A.'s to which we can refer these men and which are able to give them the right kind of reception. However, in places where there are colleges and universities this same organization of Christian Endeavorers has a very fruitful field for service with these men and women. In the Y. M. C. A. the man is limited in his acquaintanceship to men; and foreign students are as desirous of social life as we are. Therefore, an organization where they can meet Christian men and women is of vast importance in helping them to enjoy their visit to our land.

Our committee gives information and advice, and corresponds with students before they come to this country. It advances knowledge of the ways of living in this country, differences in habits and customs. It also advises about the changes in climate so that the student will not come unaware of the conditions which he will find. When he arrives in New York, or at any other port, if possible we send a representative who speaks his own language to meet him. If there is no one to help them, they are swindled by taxi drivers and by other men who are very unscrupulous in their dealings with foreigners.

We also advise by telegram a secretary of the school or university to which they are going so that there will be someone to meet them at the train, to help them to find a boarding house, and to get them registered properly in the university.

A very beautiful incident occurred at the Northfield Y. M. C. A. conference this summer which impressed upon me the great importance of this work. A handsome young Chinese got up to speak about the work of the Committee on Friendly Relations among Foreign Students, and looking heavenward with a most beautiful expression on his face he said in part as follows: "We Chinese cannot understand the extreme unselfishness of the Americans. First you send missionaries to China. You started schools there and you do not ask for any payment in return. And then we Chinese come to this country to study and you send a representative to meet us at the boat and he takes us to our colleges and then in the winter time he comes around to visit us to see how we are getting along. We cannot understand such things. There must be something behind it and we appreciate it more than we can say." It was a great inspiration to watch his face as he cited the various things which were being done for Chinese students and his very talk showed how much they appreciated it.

Now these are the ways in which Christian Endeavor societies can help. If you are not aware of the presence of foreign students in your community, upon receiving a postal card from you our office will be glad to forward the names of students in your city and with that in hand you will be able to ferret them out.

Something which we remind you of is that before any religious work is attempted, as I am sure has been suggested throughout this conference, the best avenue to personal work is through service. A kindness done to a man will throw him off his guard and make him open for anything which you may desire to do in a religious way afterward. My first suggestion is that you invite the foreign students to the socials of your society, and, secondly, that you invite them to attend religious meetings. Be sure to introduce them to your friends and to realize that they are rather bashful, sometimes because they do not understand English, and sometimes because their skin is different from yours and they think probably that there is prejudice. Try to understand that as a general rule they are accustomed to a different system of gentility from ours and little politenesses which show your interest in them will be greatly appreciated. And, in the third place, something which is very important is the inviting of these men and women to private homes for evening socials or for meals. In this way you can show him what true Christian American home life is like. A Christian influence most of them do not enjoy in their own country.

But, above all, in all your dealings with them, remember that they have souls, feelings, and difficulties, that they are confronted with things which they do not understand, just as we are, and any little kindness which you may do for them will accomplish wonders in proving that there is something to our Christianity after all.

Lastly, you may wonder what the opportunity is for definite Christian service. I will relate one incident which proves to my mind the opportunities for bringing these men and women to Christ.

A very fine young man, a Brazilian, called in at the office for some information about a place where he could buy a book in Portuguese. We telephoned to four or five places. Afterwards he asked me if I could not help him to get his bank account straightened out. I went to a bank with him and as the arrangements were not satisfactory, I took him to another. On the way back we stopped in at the Brazilian consulate to see if there was any mail for him. As we were coming out of the consulate, having been disappointed in his quest, he made a rather slurring statement about the condition of the consulate and told me the following story:

"The second day after my arrival from Brazil I went to the consulate thinking that because I was a Brazilian the consul would be very willing to help me get settled and to give me information about schools and colleges. While at the consulate I asked if I might not use the telephone. The clerk very gruffly told me that there was a pay telephone downstairs and I could use it." Then turning to me he said: "I go to your Committee on Friendly Relations among Foreign Students. I am not a member of the Y. M. C. A. I am a Catholic. I have absolutely no connection with your work at all and I ask for a little information and you give me help. I ask for a little help and you do two or three things which I have not requested. In truth you have gone the second mile to please me. This, in truth has captured my friendship."

Then on the subway I noticed that he was very quiet and thoughtful. He finally turned to me and said: "Where can I find a Bible?" He knew enough of our organization and what the "C" in the Y. M. C. A. stands for and he was willing to go as far down as necessary in his investigation to discover the power behind the scenes which enables us to do the work which we are carrying on. I am glad to say that he bought a Bible and is studying it very diligently at the present time and I hope that in a few months he will become a thoroughly grounded Christian man, as he undoubtedly is the type of man who will return as a leader to Brazil. Now that happened without a word on my part about things religious, or about the motives which are behind our work, simply through service which was rendered him, very humble in its quality, just ordinary kindnesses which anyone would render to a friend, and yet that was the result. I think that this should ring out as a tremendous challenge to every one who is interested in the welfare of our foreign guests.

The great tendency and danger is for these men to come and to return worse than when they came and unless we come in contact with them, and realize our opportunities, we are going to fail in this tremendous work which is presented to us. These men are going to be the leaders. Now shall we let them return as Christian leaders, or Bolshevik leaders like Lenine and Trotsky? This is a question which we have to decide.

I sincerely hope that we can count on your co-operation in forwarding this movement. If at any time you are not able to give the help necessary, kindly refer these men to our committee.

Are we going to allow these students to return home with all their dreams of a Christian United States shattered, or are we going to allow them to return with their faith in God torn to pieces? Are we going to allow their belief in Christ as the Savior and Redeemer to be lost? No. The opportunity is great. The helpers are many if they but respond to the challenge. The results and recompenses will be without number and I am sure our efforts for our foreign guests will be for the uplift of the world and will redound to the glory of Christ.

WHAT THE COMMUNITY HAS A RIGHT TO EXPECT FROM THE CHURCH.

By Ira Landrith, D.D., LL.D.

The church of God is God's organized omnipotence on this earth. The church therefore could do what it would if it would do what it could. But the church of God at its weakest and worst is better than any other institution at its strongest and best. Not ideal, because we have this treasure in earthen vessels, the church is still the most nearly ideal organization in existence; and without it no home would be safe and no community endurable.

The cheapest wit and the most contemptible twaddle is that which puny platform and press puppets fling at the church from their inane brain vacuums. It is so easy and safe that intellectual indolence delights in it, common cowardice approaches the pastime unafraid, and iniquity and cupidity make of it a revel and an orgy.

Yet without the presence of the church no community would be safe for home, nor children, nor character. Without the influence of the church in America this nation would be more conscienceless and sensual than were the decadent nations of that ancient civilization that now is dual. But for those "gentlemen unafraid," the ministers of local churches, and their brave and often tremendously sacrificial advocacy of right and denunciation of wrong and of greed, the brothel would be to-day the protected child of law, the gambling hell a legalized place of popular resort, the lottery a continuing partner of the State, and the saloon an honored department of debauched government. All honor to every other aiding agency, but it was the preacher in his own pulpit that created the atmosphere in which legalized crime cannot exist; in which lawlessness cannot be ignored; and in which the oppression of the poor and the racially weak and unfortunate will never be attempted by any man or organization, however rich or powerful or politically influential. Verily, the church of God *is* God's organized omnipotence on earth, and he who would serve his kind must serve God through the church, unless he would weaken his own usefulness.

Better a perfunctory preacher than no pulpit at all. But perfunctory preaching is almost blasphemy, when the world is crashing and maps are collapsing and men are tragically dying by tens of millions.

Better a conventional church of pious commonplace than a shut-up temple of the Lord; but, when unprecedented work is to be done, God pity the church that is not a modern and busy workshop.

Better a silent, up-reaching church spire than that no finger at all point heavenward. The very presence of a church building in a community is a challenge to prayer and service.

The complacent ecclesiasticism that, because truth is changeless, practises also the method of a century ago will still do great good, but its most partisan friend, if intelligent, could hardly hope that it would overtake twentieth century opportunity, that both comes and goes, by steam and electricity.

THE RIGHT-MINDED COMMUNITY.

Has a right to expect of the church—

1. Salvation of souls.
2. Salvation of the community.
3. Anything and everything that helps———.
4. Vital faith.
5. Co-operation.
6. Conservation and economy.

THE WRONG-MINDED COMMUNITY.

Has a right to expect—

1. No compromise with public or private wrong.
2. Unafraid antagonism of evil.
3. An atmosphere in which legalized iniquity cannot survive.
4. And all in the spirit of Him who wept over a wayward city about to be destroyed.

THE CHURCH OF TO-MORROW.

Will be—

1. More, much more than a club.
2. More than a place of rest.
3. More than the home of the fortunate.
4. More than a forum for "safe and sane" propaganda.

THE SUNDAY SCHOOL AND THE YOUNG PEOPLE'S SOCIETY.

By Rev. Frank Otis Erb, Ph.D., Philadelphia, Pa.

One of my most interesting recollections is the memory of seeing the Siamese twins for the first time. Here were two persons joined together by living tissue. They each had perfect heads and bodies and arms and legs. They differed in complexion and temperament. But the same life-blood flowed from one to the other. That which helped one helped the other, and that which injured either injured both. They might have differences of opinion. One might sometimes wish to go in one direction while the other desired to go in another. But they always ended by going together or not going at all.

The relation of the Sunday school and the Young People's Society is best illustrated by the Siamese twins. Or, rather, it is a case of Siamese triplets or quadruplets or quintuplets or worse. For in your churches we have not only a Young People's Society and a Sunday school, but we have also such organizations of young people as mission circles and guilds, young men's clubs, Boy Scouts, Campfire Girls, and so on. In a very real sense these all live by the same life. Not only do the membership and plans of these organizations overlap, but the general life of the church largely determines their vigor, vision and achievement. If the church is progressive and aggressive, these societies are also alive. But if the church is decadent, the affiliated organizations also live a struggling and starving life.

It becomes a self-evident truth that rivalry and strife and jealousy are not only out of place, but suicidal. The Packard Piano Company have as their motto, "If there is no harmony in the plant, there is none in the piano." In every sort of organization this fact is obvious, above all in church work, where motive and spirit and purpose count so largely. Even to raise questions of primary and secondary, superior and inferior, is of doubtful worth, since from some points of view each is primary and each secondary. In the old days when the common mode of locomotion was by horseback, Dryden was undoubtedly right when he said, "If two men ride on a horse, one must ride behind." But we live in an automobile age, and in a Buffalo Pierce Arrow car, three can ride abreast more comfortably than singly.

Neither do we solve the problem of the relation of the Sunday school to the Young People's Society by proposing that one swallow the other. This solution might have been correct in jungle days, when

"A lady who lived on the Niger
Went to ride on the back of a tiger.
They returned from the ride,
With the lady inside,
And a smile on the face of the tiger."

But in civilized days this solution, though apparently easy, is really most difficult. It will ultimately raise more questions than it answers. The fish that swallowed Jonah discovered that a worth-while thing is not to be swallowed with impunity, and in any case is not to be kept down.

The fact is, of course, that there are great values on each side, values that must not be sacrificed. On the side of the Sunday school are its comprehensive membership, its improving plans for Bible study, its adult leadership, and its widening vision of worship, education, and service. On the side of the Young People's Society are its splendid enthusiasm, its spirit of devotion and confession, its recognition of the place of service and recreation in an all-around life, and its training in democratic living.

But both sides have their weaknesses as well. The Sunday school is woefully deficient in its worship. It is even more deficient as a democratic institution. When the Sunday school, as is ordinarly the case, is "run" by the superintendent and a few others, it is as autocratic as the Kaiser. In an age like ours, when the whole world is becoming democratic, the young people's section of the Sunday school must become a training-school in democracy.

The Young People's Society, on the other hand, has its weaknesses. We have not yet learned how to graduate folks who fail to realize the truth of the old song,

> "The old gray mare, she ain't what she used to be,
> Many long years ago."

The only ways we know to graduate them are those of the undertaker or the Bolshevist: we must either bury the society and start another or wage "Red" revolution against it.

Nor has the society learned to utilize adult leadership. One of the great needs in most of our organizations is for a few older people who will be neither bosses nor office-holders, but who will be able graciously to spur or check or guide, as the case may be, but always to inspire.

The main weakness is, however, that neither the Sunday school nor the Young People's Society has discovered how to merge its life into the life of the church, and develop a programme that will meet all the needs of all the young people, and at the same time will challenge each organization to make in largest measure its peculiar contribution to the life of the whole.

In many places, they are working out the "Council Plan" to accomplish this. The Council is composed of two representatives from each organized group of young people, and also includes the pastor, the Sunday school superintendent, and one or two other adults who are sympathetic and inspiring in their relation to youth. The young people, however, fill most of the offices, and assume responsibility for the plans and methods adopted. The Council plans for all the life of all the young people, assigning to particular organizations those functions in closest agreement with their genius. Thus the Sunday school becomes mainly responsible for Bible-study whether on Sunday or week-days, and also for teacher-training. The mission circle becomes mainly responsible for the diffusion of missionary knowledge and enthusiasm, whether by study-classes, meetings, pageants, or literature. The Young People's Society assumes as its function the devotional life of the young people, as represented by the personal devotional life of its members, prayer meetings, cottage meetings, and personal-work classes. To the pastor belong evangelism and the challenge to life service. Other needs, such as education, service, and social life, are assigned to special committees raised for the purpose.

When our allies abandoned their national plans for a world programme and unified leadership, they drove the Hun from the sea, from the air, from France, from Italy, from the Balkans, from Syria, and from Mesopotamia—drove him back to Berlin. When we learn how to unify the forces in our local churches, we too shall go from victory to victory.

CONFERENCE ON ALUMNI ACTIVITIES. "CLEVELAND'S EXPERIENCE."

By Mr. Fred L. Ball, Cleveland, Ohio.

In March, 1916, while returning from a trip in Northwest Canada, Dr. Poling stopped in Cleveland long enough to meet a few of the workers

Another View of the Canadian Rally.

Dr. Daniel A. Poling,
Speaking at the Canadian Rally.

of the Cleveland union, and at this meeting he suggested an organization that would promote Alumni Fellowship and also assist in a financial way the programme of the United Society of Christian Endeavor. His devoted interest in the cause of Christian Endeavor has long been known to his many Cleveland friends, and it made a new impression on the minds of those who attended the meeting. A temporary organization was effected to arrange a suitable opportunity to present the proposition to as many friends of the movement as the commitee could get together. The meeting was held May 23, 1916. A supper attended by about 100 and an after-meeting of about 300 were held. In characteristic style Dr. Poling presented the opportunity of a wider fellowship among those who had served and were benefited by Christian Endeavor. The immediate need and the definiteness of the appeal *won* in the hearts of 76 of those present. An organization was formed, and it was decided to call the Cleveland Fellowship the Poling Foundation. This name has since been changed to "The Poling Foundation Alumni Association," conforming to the plan announced by the United Society. $463.75 was pledged each year for five years at this first meeting. Dr. Poling was pleased beyond measure, and announced that the funds would be used as wisely as possibly, and a *definite* and *detailed* report would be made to the Foundation upon his next visit to Cleveland. The next meeting of the Foundation was held about one year later, April 13, 1917, when 85 new members joined and 19 of the old members increased their pledges.

In all, six meetings have been held when Dr. Poling has been present. Dr. Poling deserves all the credit for any success the Cleveland Fellowship has attained. His reports of what has actually been done with the funds contributed has been one of the greatest helps in retaining the interest of present members and enabling us to secure new friends in the Fellowship.

Under the heading of plans, I am glad to say, we had only a very few. They were simple. We organized to get a list of honorary members from each society. Invitations were sent to this list. Personal following up helped to get the interest of some who had been out of active touch with the work for even a few years. Such names were given to members of the committee known to have an influence. Many of our most substantial givers were secured by this plan. We are organized now under the plan of the United Society. We meet about once or twice a year. We purpose to have only one financial appeal during the year. Other meetings are simply for fellowship.

Results of work may be summarized as follows: Present membership, 467. (Only 8 have withdrawn memberships.) Increased pledges received from 99. Amount paid to the United Society, 1916, 1917, 1918, $3,000.

Amounts due from pledges for the next five years: 1919, $2,351, plus $200 in Liberty Bonds; 1920, $2,316; 1921, $2,125; 1922, $1,976; 1923, $1,-429.50; total, $10,197.50; plus Liberty Bonds, $10,397.50.

STEWARDSHIP, THE NEXT GREAT REVIVAL OF THE CHURCH.

BY DR. WILLIAM HIRAM FOULKES, *General Secretary, The New Era Movement of the Presbyterian Church.*

(NOTE.—Dr. Foulkes was not able to be present at the Conference. However, we publish the address he intended to deliver.)

I shall begin my message with five barley loaves, or the miracle of a boy's heart. It was a likely, lovable lad who said to his mother on that memorable morning:

"Mother, please let me follow the Master to-day. He is so wonderful and so kind. He does not talk like the synagogue teachers. He is always telling us about the flowers and the stars, about our Father and His children."

And the mother said,

"But you never know enough to come home when you follow the Master. You walk and walk until you get all tired out."

"Please, mother, let me follow the Master to-day. Please!"

And then the mother with a little show of impatience, and yet with real affection in her heart, said,

"Well, there, run along," and as an afterthought added, "But you must take a little lunch with you, for you will get hungry before night....... I'm sorry I haven't much to give you, for we are, oh, so poor;...... but you can have just the best we have."

And thereupon she gave him five crackers and two sardines. I know that the story of the boy tells us that he had "five barley loaves and two small fishes." We must not think of great big baker's loaves or of any kind of large fish. I suppose they were really dried herring, but we can interpret the story into the language of to-day and say "sardines," or as I used to think when I lived out on the Pacific Coast, "smelt." Then the barley loaves were only cheap, common, flat crackers. But they were all he had.

Then came the incident around which my message is to centre. The Master stands in the midst of a troubled group of disciples facing a great crowd of people. Thousands and thousands of them, like the boy, have been following Him all day. They are tired, hot, and hungry. What will they do? What shall He do? He shares His concern with the disciples.

"Whence shall we buy bread that these may eat?" He asked them, not because He did not know, but because He wanted to put His disciples to a great test—the test of stewardship. It is an interesting answer that the disciples made to His question. It is as old as human nature and as common as selfishness: "Send them away."

That has been the answer of millions of professed followers of the Master ever since, who, facing the famished folk of earth are always saying, "Send them away!"

The Master has His answer for their answer, His command for their indifference: "Give ye them to eat!"

He riddled their unbelief, their selfishness, and their unconcern with His mighty imperative "Give!"

How could they? They had scarcely enough for themselves. They had not money enough to buy, and there was no place near enough to furnish them a full supply. They had worked out the problem with their human arithmetic, and found it hopeless. It could not be solved.

Then the boy appeared with a key which unlocked the door. It was his five barley loaves. Probably all the time he was as near the Master as he could get. Trust a boy to be near the celebrity. He was doubtless near enough to hear the dialogue between the Master and His disciples; to see the look of entreaty upon His face and of perplexity and dismay upon theirs.

All at once, the great miracle took place. Where? In a boy's heart! He wanted the Master to have what he had. He wanted to give the things which had been given to him, to one who had a rightful claim upon his life. Surely the boy did not dream the result of his giving. He did not know that his crackers and sardines would feed the thousands. He only knew that the Master needed something, and that it was his privilege to give the Master what he had.

That was the heart of the ancient miracle; it is also the heart of the modern miracle. It is the key that will unlock the door of the greatest revival the church has ever known. When the church learns the lesson of the boy and the five barley loaves, it will be able to teach the whole world the real meaning of life.

A new conception of the meaning and power of stewardship has entered into the thought of the Christian church. The world's unrest and the recent world war have compelled the church to analyze its attitude toward things. Are "things in the saddle" forever "to ride mankind"? Is the Christian to be carried away by what Walter Pater calls "the tyrannous reality of things visible," or is he to recognize the true doctrine of "the thing"? Our attitude toward things will indicate our real relationship to the realities underneath things. Human possession is a pagan dogma. Human stewardship is a Christian doctrine. There is only one owner, God. The Christian is a trustee. He is God's possession, bought with a price.

If this truth seems to rob him of the sense of ownership, it adds to him the joyous responsibility of stewardship. All that one has, if he be a Christian, is under the obligation of trust. His stewardship also involves a recognition of this threefold relationship in his attitude toward things. He sustains an attitude toward God, toward his fellow men, and towards his own life. What he does with things will show what his real attitude is toward his God and his fellow men. If he wantonly wastes, or, on the other hand, niggardly withholds the things which God has given him, he is not a good steward.

The Christian's stewardship also involves setting apart a definite proportion of his income, to be devoted exclusively to others. For the good of his fellows and the good of his own soul he must "take his own hand off" a certain definite share of the things he has acquired in order that the real Christian balance may be preserved.

What a revival would come and will come when every follower of Christ sets apart a definite, worthy portion of his or her income! There is already money enough in Christian purses, if Christ's rightful share of it were devoted to Him, to build every needed hospital, school, and college; to equip every individual church with men and property; to reach every unevangelized community with the full-orbed gospel, and to bring in His kingdom in a day.

The great war-drives and loan-appeals have demonstrated the volume of giving which is possible when every one gives a little. What a mighty stream of power would flow forth from the church if every one of its members were to give, not a little, but that full share which is God's right! If our dull ears could only be unstopped to hear the clarion call of Christ voiced through the urgency and immensity of human need, "Give ye them to eat!" With such a revival there could never again come to pass unchecked famine, and a hundred other unrelieved human ills to which we have become so accustomed that our hearts are calloused.

It is too late to begin with the generation that is now in its zenith. If this mighty revival is ever to come in, it must be carried on the crest of the wave of youth. The delegates to this Conference have more power to accomplish the impossible and to lift the giving of the church from the plane of beggary than any gathering of similar size, of bishops or deacons or preachers.

The widening influence of the Tenth Legion and the inbuilding of the principles of Christian stewardship throughtout the whole Christian Endeavor world will go farther than any other single thing of which one can judge to enable God to usher in His kingdom of righteousness.

The world to-day is hungry for the bread of life; for kindness and brotherhood; for justice and righteousness; for mercy and truth. Shall we Endeavorers repeat the apostolic mistake of trying to "send them away that they may buy bread for themselves"? Rather let us perform the miracle of the barley loaves in our own hearts first of all, and to devote to Him all that we have and are. With such devotion on our part, not alone will the thousands but the millions be fed. It is not five barley loaves that the church of Christ in America has to offer, but the harvest of a billion bushels of wheat; of wealth that has poured into our national coffers from every land on earth; of riches in things and in stored-up service. Expenditures are upon a vaster scale than ever before in the history of the world. We speak of billions when we formerly mentioned millions. Wages as well as the cost of living have gone up in horizontal scale. Neither the nation nor the church can endure this increase of opulence unless they recognize the eternal obligations of stewardship. It is not what America hoards but what she gives that will save her and the world.

Far above all the glory of the stewardship of substance is the stewardship of life. After all, Christ has waited for nearly two thousand years for a generation of people who will give themselves as utterly to Him and His needy world as our sons gave themselves to their nation's honor and the need of the world. What priceless memories are ours from Flanders fields, from Soissons, and Belleau Woods! Shall our brothers and our sons, our lovers and our friends, have died in vain? No, if we who bear the name of Christ match their heroic devotion with our own, and place our lives upon the altar of sacrifice, a whole burnt-offering.

There are enough young people here to-day, with barley loaves in their hands, to feed the famine-stricken folk of earth with the bread that cometh down from heaven, if they will only, but absolutely, yield themselves a living sacrifice, and in the words of the sainted Havergal pray,

"Take my love, my Lord; I pour
At Thy feet its treasure-store;
Take myself, and I will be,
Ever, only, all for Thee!"

LATIN AMERICA'S PLACE IN THE WORLD'S LIFE.

By Samuel Guy Inman, New York City, N. Y.

With the signing of peace the world's attention will shift from Europe to Latin America. Food, room for overcrowded populations, and a market for surplus goods and capital are the great quests of the world, which will compel the wooing of these twenty great, rich, undeveloped nations to the south.

Beginning at the Rio Grande and stretching down through Mexico, Central America, across Panama, over Venezuela, Columbia, Brazil, Peru, and the abounding plains of Argentina to the Straits of Magellan is the largest stretch of undeveloped fertile land in the entire world. All the population of the world could fine place here and be only one-third as crowded as is Porto Rico. Argentina, far more capable than New York of sustaining a dense population, would have 220,000,000 people instead of 8,000,000, if it were as densely populated as that State. This is why capitalists, manufacturers, steamship directors, food economists and political leaders in Europe, North America and even Japan, are so intently fixing their attention on these fallow lands. A great part of our newly built merchant marine will be assigned to the West Indies and Central and South America. The biggest packing company of the United States recently shifted its most important organizer from New York to Buenos Aires.

The world war has led Latin America to open wider her doors to the United States. Stopped by the war from getting their capital and manufactured goods from Europe, these countries suddenly appealed to us, and the Pan-American Financial Congress in Washington was our Government's answer. To-day, North American banks, ships, and stores are seen everywhere. These have helped break down the strong prejudice that has existed against the United States. But what had done still more to bring them to understand and appreciate this country has been our entrance into the world war. The "Yankee dollar chaser," with his Monroe Doctrine of "America for North Americans," has changed into the unselfish defender of democracy, renouncing profits in munitions to fight the battles of democracy for every American nation. So now, instead of the old warnings, "American solidarity" is the universally expressed desire.

The new spirit of confidence is demonstrated by the increasing number of Latin-American students coming to the United States. Not long ago, twenty-seven arrived on one ship from Brazil, all sent by the national government, which government requested that the student be met and helped to their schools by the Christian forces here that supported the missionary work they had so approvingly witnessed in Brazil.

A law has been introduced into the national Congress to provide several scholarships for North American students in Brazilian colleges, and to endow a chair of Brazilian history in a North American university. The same government has recently employed a young woman missionary to become principal of a large Girls' School for Domestic Science and Home Training and sent her to the United States to secure seven other young women to help in the project, prescribing that all of them shall have the same missionary spirit that she has.

The world war has shown Latin America her need of religion. Her intellectual classes, becoming disgusted with the old formal religion, opposed to all progress, had come to think that science, philosophy, and economics were sufficient. But it is doubtful if anywhere in the world the crash of the world's most scientific nation has been more keenly felt than in Latin America. An illustration of the way leaders are awakening to the need of a spiritual basis for national and personal life is the recent suggestion of a university professor of Buenos Aires that a Congress of Religion be called to find a basis for religious life before his nation was lost entirely in materialism.

Many national leaders are turning to North American missionary forces to direct either new institutions or old ones which, while well equipped, have failed because of lack of unselfish, devoted service on the part of the directors. Often there is frank acknowledgment that they have not the altruistic prepared leadership which they realize can be found only among North American evangelical Christians.

The new day is well expressed by a word just received from a leading missionary of South America, who has an open door in practically every university of South America:

"With the present spiritual unrest that signifies a deep longing for something morally and spiritually better, and with the United States standing to-day beside France in the affections of the South American peoples, one longs to see every North American agency that can make genuine contribution to the moral and spiritual progress of South America give itself wholeheartedly to this opportunity for Christian effort in South America. No such time has existed since the days following the gaining of their political independence."

Fortunately, the Christian forces of North America are prepared to take advantage of this opportunity. Following the Panama Congress in 1916, thirty mission boards have united in a board of strategy known as the Committee on Co-operation in Latin America. Territorial responsibility has been fixed so as largely to eliminate overlapping, and inclusive programmes have been mapped out for this continent and a half. Union colleges or theological seminaries have been recently begun in Mexico City, Porto Rico, Chile, Argentina, and Brazil. An international union "Faculty of Theology and Sociel Sciences" is soon to be opened in Montevideo, the Hague of South America, where graduate training will be given to students from the whole continent. A comprehensive programme for the production of an extensive literature in Spanish and Portuguese is being rapidly developed, with all the mission boards co-operating. These large, comprehensive plans call for leadership of the first class and offer positions of influence that will reach out over a continent.

The cry of to-day is no longer Horace Greeley's "Young man, go West," but Ezekiel's "Son of man, set thy face toward the south." There to the south you will find rich young nations, facing a great future. There you will find the forces of the old world battling for cultural and commercial supremacy. There men and women of deep conviction, of living Christian experience, of loving sympathy and of ample preparation, will find open to them avenues of influence that sweep across a continent of opportunity.

AMERICANIZATION OF THE FOREIGN BORN—A LITTLE STUDY IN CHRISTIAN CITIZENSHIP.

By H. F. Shupe, D.D., *Editor of the Watchword.*

Christianization would be better, but the "Americanization" of the multitudes who come to our land is essential to their welfare and to the safety of America, and deserves our full attention. Its significance appears when we contrast "Americanize" with "Russianize" or "Prussianize."

America as a melting-pot has been tested by the war. The fusion produced in its fires has been found defective. The strength of America is threatened.

A screen is needed for the melting-pot. That screen or sieve is now in process of making in the United States Congress, and Christian citizens should lend their wisdom and encouragement to the end that the screen may be effective and fair, lest in keeping out the raw ores some valuable metals may not unnecessarily be excluded.

The first step in the Americanization of the foreign born is the *selection of the material,* or rather the exclusion of the elements that cannot easily be amalgamated. The anarchistic, which is opposed to government; the defective and diseased; and the criminal should be excluded by rigid immigration laws. But *quantity* also enters into the formula for the making of the right kind of American citizens out of the aliens—perhaps we should call them our allies. The immigration-law screen, therefore, should admit only such numbers as the melting-pot can fuse safely.

The Committee on Immigration and Naturalization of the House of Representatives has had under consideration the *percentage principle for regulating immigration,* a principle indorsed by the Federal Council of Churches. Briefly stated, it is a plan by which immigrants of a given race would be admitted on the basis of a certain per cent. of the number of the same race who have become American citizens. For example, in 1914 about 324,000 Italians came to the United States. If a percentage law had existed, limiting the Italian immigrants to five per cent. of the Italians already American citizens by naturalization and birth, the number would have been reduced by 150,000, and only 45,000 men would have come in. This process applied to all the other tongues would greatly limit immigration of people who are slow to become Americanized. This process continued would bring to bear upon the problem the influence of the Americanized people of every race upon their own people.

Other features of this proposed immigration-law screen are an effective immigration commission, higher standards of naturalization, and the application of this law to all races, thus removing the stigma of broken treaties with China and Japan and eliminating a cause for war.

Christian citizens should inform themselves as to these features and vigorously aid the House Committee in putting into the law scientific and righteous principles of immigration control.

A member of the House Committee on Immigration expresses the fear that selfish interests will not hesitate to give offence to other nations "simply because our boys have demonstrated that they are able to fight."

The fires that burn in the melting-pot must be intensified. The nation that has shown its ability to create national sentiment and develop loyalty must use its powers in carrying to the foreign settlements the spirit and principles of intelligent democracy. An Immigration Commission should direct in the distribution of immigrants on their arrival, and should carry to them the American spirit.

The nation should use its public schools for applying the processes of assimilating the foreign-born and foreign-speaking youth. It is significant that the American Federation of Labor favors the expenditure of enough money to make the public schools more helpful to those who work, by providing vocational training, the teaching of English to the non-English, and the preparation of the people for citizenship.

But to the *church* remains the obligation to give the term "Americanization" its *Christian content.* The unity of the human race as it came from God, who "made of one every nation of men"; the elimination of race prejudice in the Christian spirit, where there can be "neither Greek nor Jew"; the cultivation of neighborliness as in the Christian community must feature the Christian Americanizing of the strangers who come to our land, and the church of God must furnish these elements. Here Christian Endeavor can function.

"Hunk Town," on the West Side, was long ignored by the churches, but it grew apace, presently crowding the American community. The growing sense of responsibility in the Protestant churches was retarded by the establishing of a Protestant Hungarian church. But the Catholic and the Protestant churches both ministered in the Hungarian language, and there was little perceptible Americanization among these people. The mothers always are the last to be reached by the ordinary processes; the men work where they must learn some English; the children are on the streets and go to school; the mothers remain foreigners in language and habits of thought. At length the women of the American churches acted. A storeroom was secured and made attractive. A Hungarian woman, American born, was put in charge of the work for the women of the settlement. The home life was touched and Americanization was carried on in the realm of the Christian spirit.

In this work the Endeavorers of the city found a field for usefulness, visiting the homes, caring for play activities of the children, teaching English and Bible classes, etc. Personal touch in the name of Christ and in the Spirit of Christ helped to develop a degree of Christian Americanism.

A company of Endeavorers one day in July went with Dr. W. H. Morse, of Hartford, Connecticut, to New York City. There they watched a big liner depart from a Hoboken pier. It was an Italian boat. In reply to a question, a uniformed man at the wharf said, "Four thousand, three hundred and seventy Italians going home."

"Why do they go in such crowds?"

"For four years they have not had a chance to visit the home folks, and now they are going."

Ethelbert Stewart, of the Department of Labor, reports that a million, three hundred thousand aliens will go back to the old country as soon as they can get transportation, and that these men will carry with them an average of $3,000 each—almost $4,000,000,000.

As Doctor Morse and his Endeavorers watched the Italian liner with its more than 4,000 men depart for Italy, one of the girls said,

"Of all sad words of tongue or pen,
The saddest are these, 'It might have been.'"

"What might have been?"

"The Young People's societies might have furnished brothers and sisters to these Italians; they might have shown an interest in Columbus Day; they might have gotten more into the night-school; they might have gotten more of them into the missions; they might have given them Bibles —they are going back to Italy now, where it is against the policy of the Catholic Church to let the people have the Bible in their own language. Who knows what a lot of good might have been done if they were taking Bibles back with them?"

The Swede officer standing near quoted a Swedish saying to the young people, "Might be(e)s never swarm, but May be(e)s do."

A GOOD PRAYER MEETING.

By John F. Cowan, D.D., Pittsburgh, Pa.

The way to get a good prayer meeting is to buy it, paying highest market price, cash down, no credit for good intentions, no old automobile tires or geese-feathers taken in exchange. The 120 in the upper chamber worked ten days to pay for one pentecostal prayer meeting.

The price of a good prayer meeting, like that of any victory, whether in France or in the Jonesville church, is *blood;* not always blood spilled, but forever and eternally blood *spent.* It is heart-beats, brain-throbs, tissue and muscle worn out in service. Work is blood glorified. The secret of a good prayer meeting is the Spirit of God honoring honest, self-sacrificing endeavor. You can't think of a good prayer meeting except in terms of hard work. Somebody must toil and sweat to prepare for it and bring it to pass. Some one must give painstaking, laborious attention to such details as planning and writing out a programme, selecting the best hymns, mastering the thought of the Scripture lesson, and elaborating such other details as sending invitations, reminding the thoughtless by telephone or calls, decorating and ventilating the room, maybe dusting chairs and cleaning lamp-chimneys, oiling the squeaky organ-pedal, arranging seats to the best advantage, giving cordial welcome to strangers, securing participation of the backward, leading up to the spiritual climax and the evangelistic aftermath.

All these must be worked over as painstakingly and perspiringly as a good salesman thinks over and prepares his sample display, his approach to his customers, his dress, his smile, his tone, his introductory words, his finale, in order to put over the sale of a $1,000 bill of goods.

Mind you, a good prayer meeting is just good salesmanship for the kingdom of heaven; and not until leaders study and plan, as unsparing of themselves as keen American commercial salesmen toil to master the principles of psychology and the practice of scientific salesmanship, will the prayer meeting be 100 per cent. efficient and as popular as the "57 varieties."

So many of us Christian salesmen think we have nothing more to learn When a commercial salesman reaches that point he is called a "boob." A boob is a bonehead who is hermetically sealed against new ideas. Do you know, some of the newest and best ideas are the oldest, just as some of the wittiest modern newspaper jokes are being dug up by archæologists in the records of Babylon. The "mother-in-law" joke is 6,000 years old.

Now, I am going to hand you three requisites of a good prayer meeting: The first requisite is *prayer.* The second requisite is *prayer.* You know what the third is. Mind, I am not saying "prayers," but *prayer.* You can't gauge a prayer meeting with a counting machine; you have got to use the thermostat. Prayer isn't just words; it is a *spirit.* It is Spirit answering to spirit, as the Swiss yodler sings his good-morning to his waiting beloved, and the Alps send back her response.

Let me tell you, whoever else attends your prayer meeting, it will be a failure with "Mr. X. Peck Tation." "The very expectation, the looking forward to the Saviour's, "There am I in the midst," is the primal element of a good prayer meeting. When all assemble with expectation, then the girl at the organ or piano makes the instrument pray. Some organists make it do a one-step; some make it yawn and snore. Prepare your organist to make her instrument vibrate with prayer.

A brief illuminating explanation of the hymn, a glimpse into the soul of the writer, may make the difference between singing like phonographs or like a heavenly chorus.

The reading of the Scripture may be a stilted elocutionary performance, a treat-mill grind, or a spiritual impulse that will fill the room with prayer and make every heart vibrate in tune with the Infinite.

Even the necessary announcements may be prayer. I am sure that when Peter stood up in that upper-chamber prayer meeting to announce that the nominating committee would report, and balloting begin for a successor to Judas, it threw no damper on the prayer meeting. As the lightning flash announces the coming storm, every heart hushed in expectancy of the Spirit's benediction on that bit of unfinished business. You say you can't have a spiritual prayer meeting and business meeting together; was there ever a greater height of spiritual power than the pentecostal climax of that combined business and prayer meeting, the announcement as well as the prayers of which were "heard 'round the world"?

Why, you can even make the offering p-r-a-y instead of p-r-e-y, if you get into your giving Paul's spirit of "hilarious giving." Some of the most spiritual meetings I have attended this summer have been those in which our Methodist and Presbyterian and other friends were "going over the top" in their great drives for "millions for missions."

And so the prayer meeting that is planned for with endeavor and prayer, that is entered into with keen expectation, will move on through all its periods with the spirit of prayer in everything, and on an ascending scale that leads to climax. You wouldn't think of counting words, but you would count the glow on faces of those going home singing in their hearts because they had met the King of kings, and because He would abide with them until the church chimes again call:

"Sweet hour of prayer, sweet hour of prayer,
That calls me from a world of care,
* * * * * *
And since He bids me seek His face,
Believe His word and trust His grace,
I'll cast on Him my every care,
And haste to thee, sweet hour of prayer."

COMMUNITY SERVICE.

By Rev. C. E. Hetzler, *Field Secretary West Virginia Union.*

It has been wisely said that "they also serve who only stand and wait." Considerable significance may be attached to the idea that they "stand,"—ready for service,—minute-men in the particular cause which they represent.

But even the privilege of being "minute-men" is not sufficient for warm-blooded Endeavorers, whose hardest task is to wait. It is contrary to their natures, and why should they stand idly by when so much real service awaits them at their very door? Young men and women of ambition and caliber want to make their lives count for the most, and this is attained by attending to the duties at hand, whether great or small. Many of them will, indeed, seem small and insignificant, but may we be reminded of the little memory gem of our childhood:

"Little drops of water; little grains of sand;
Make the mighty ocean and the pleasant land."

We are conscious of the fact that the world is fast becoming a community; but this vision does not, and indeed never will, abrogate, the needs and opportunities of service in the immediate community. Moreover, he who serves the immediate community serves the world community, and world betterment is hindered according as community betterment is neglected. He who aspires to world service must prove himself in community service.

Those who seek opportunity for community service should investigate the needs of the community, for every need constitutes an opportunity for some one, and every opportunity is a call to some one, in the name of Him who said, "inasmuch as ye did it unto one of the least of these, ye did it unto me."

While we may have been disappointed with the number of "enlistments" in the Universal Army of Patriotic Service, let us console ourselves with the consciousness that the shortage consisted simply in the neglect of signing the name, for Endeavorers were scarce indeed who did not actually do some bit of patriotic service. With the ramification of the effects of the war into every nook and corner came the consequent and corresponding challenge to the best in manhood and womanhood everywhere.

During the early days of Uncle Sam's participation in the world war a pastor in a small city in which a soldiers' camp was located conceived the idea of religious services for the soldiers. The services were designated as Christian Endeavor services under the auspices of the local Endeavor society, and a temporary organization was effected. The society was named in honor of the obliging captain, and every Sunday afternoon this pastor would take a few Endeavorers, go to the camp, and there conduct services for the boys who were not on duty. Singing, prayers, testimony, and a short talk on the current Endeavor topic featured the service. Though no great accomplishment was evident at the time, letters from the boys after they arrived in France expressed appreciation of this bit of interest in the boys' spiritual welfare, and gave evidence of results.

But the war is over, and the peace pact is signed. What about peacetime opportunities for service? The great problems of reconstruction are before us. What part will Endeavor play in this stage of the game? If we are to justify in any measure our claim that Christian Endeavor is the greatest training agency in existence for young people engaged in Christian service, we must not only be trained, but we must serve. There are many opportunities of national and world scope which confront Endeavor, and we are confident these will be met with true Endeavor spirit; but since we are to think now of "community" service, we will speak rather of the "little drops of water, and the little grains of sand."

Any effort put forth, or energy spent, in the channels of the local church of which one may be a member would contribute to community uplift. To live an exemplary life before one's neighbor, and then by a few well-chosen, heaven-blest words to lead him to Christ, would be effective personal evangelism. There is a tremendous duty for Endeavorers in the matter of community evangelism, which should become an opportunity as we enter into the joys of Christian service. Rauschenbusch is near the truth when, in referring to the commissioning of the seventy as the first great home missionary enterprise, he remarks that home missions is only returning to its own first beginnings when it begins to deal with the working masses and classes. The challenge to Endeavorers everywhere is to evangelize and Christianize the immediate community. Shop meetings, open-air services, community Bible classes, and many other such methods of visualizing Christianity to the world will do much to bridge the chasm between the church and the unchurched.

Shailer Mathews says, "The greatest service which the church can render society just at present would be to contribute the spirit of Jesus to the ideals which are provocative of discontent." A closer study of economic and sociological conditions of to-day, followed by putting into practice the plans and purposes evoked by such a survey, would commend the Christianity we advocate to those who, with or without good reason, are set against the church. If they, bound by the spirit of prejudice, shun us, let us, impelled by the spirit of Christ, seek them. A wise-spread campaign of visitation might be instituted by Endeavorers some month, during which time thousands of Endeavorers might be willing to visit unchurched homes, of rich and poor alike, for the sake of observation, looking to a broadening of their conception of conditions; also for the purpose of presenting Jesus Christ and inviting the members of the homes to church. If the mountain will not come to Mohammed, let Mohammed go to the mountain. A census summary, if there has been a recent census, would aid the efficiency of such a campaign, and if there has been no such recent census, why should not some Endeavor society or union conduct such a census as a bit of community service? Is it beyond the province of Endeavorers to start agitation on matters of reform for the community? Not the least factor in many great movements was the pioneer of faith and vision, who had courage enough to stand on the housetop and sound the trumpet.

The social life of the community furnishes a worth-while channel of service for the Endeavor society. Jesus said, "Ye are the salt of the earth" and included us. Endeavor sociability will not be a clique affair, but will savor of the spirit of the Christ who dined with publicans and sinners; who walked with fishermen by the seaside; who talked of the water of life to the woman at the well; who wept over the wicked city; who ministered to the sick, the lame, the blind, and the poor; who graced, with his presence, the wedding at Cana; who blessed the little children with his kindest attention.

We have been thinking of serving the spiritual, mental, and social needs of the community. We may serve these interests directly, as suggested, or indirectly by an appeal to the æsthetic in men and women about us. We may crowd out the evil in men's hearts and lives by substituting the good, the beautiful, and the true. Lives are ennobled by the influence of beautiful music. Chateaubriand says, "Music is the child of prayer—the companion of religion." Some Endeavor society could very well be responsible for a series of first-class musicals, a Chautauqua, Saturday afternoon athletic contests, or something of this nature, in the community.

One of the finest kinds of service which an Endeavor society could render to a community would be to conduct a garden and lawn contest. This would mean real material benefit to all who would take part; but, above all, the very "tone" of the neighborhood would be improved, because the people continually would feast their eyes, and consequently their souls, on the beautiful lawns and flowers, and unsightly places of the community would be screened from sight by the aid of nature. One man spent hours of toil, delightful to himself, in cultivation of flowers and grass, around his premises. He would spend a few moments morning and evening "with God in His garden." It meant much to his own soul, and he often said that one of the greatest blessings he received from his labors was the pleasure of seeing and hearing passers-by as they admired the lawn and flowers. Do you not think that many a man and woman went to their daily routine of work with a little more cheer in their hearts, less of evil thought in their minds, and more courage and spirit for the day because, for at least a few moments, their minds were free from contamination of sensual and mercenary thoughts, as they looked upon God in nature.

We would gladly continue, in these pleasant and profitable reflections, concerning the opportunities and privileges open to ambitious, buoyant, hopeful, consecrated, undaunted, Endeavordom, but we close with the words of Maltbie Babcock, who, though he may not have been a member of any particular Endeavor society, certainly had the true Endeavor spirit, or he could not so beautifully have voiced the secret of the virility of Christian Endeavor, which is service, in the words:

"Be strong!
We are not here to play, to dream, to drift;
We have hard work to do, and loads to lift;
Shun not the struggle—face it; 'tis God's gift.

"Be strong!
Say not the days are evil. Who's to blame?
And fold the hands and acquiesce—oh shame!
Stand up, speak out, and bravely, in God's name.

"Be strong!
It matters not how deep intrenched the wrong,
How hard the battle goes, the day how long;
Faint not—fight on! To-morrow comes the song."

ADDRESS OF WELCOME TO THE CONVENTION BY THE BUFFALO FEDERATION OF CHURCHES, AUGUST 5, 1919.

On behalf of the Federation of Churches of Buffalo, I am honored in the privilege of extending to you all, to Dr. Clark and his associates at Boston, to the hundreds of officers of the Society from many States and lands, and to the thousands of members in attendance, a most cordial and hearty welcome to Buffalo.

The Federation of Churches includes 20 denominations, various other organizations in all 124, banded together for civic righteousness, moral spirit, the restraint of vice, the defence of the Sabbath, and the maintenance of temperance. It is a mighty force in this great city for law and order.

We welcome the great Christian Endeavor Society as itself a federation for Christ and the church. Its pledge keeps it in constant loyalty to thousands of ministers, its growing membership assures the church that an army, invincible and inspiring, gathered from many lands, is the reserve power for the coming years, and its united spirit is the bright omen of a day when we shall all be one, our various colors melting into the white light of compact and indestructible truth.

We wrestle against principalities and powers, against the rulers of the darkness of this world, and we see in your bright faces turned to the east a harbinger of the coming day.

In the name of this body of thousands of Buffalo church-members, I welcome you to our homes and churches and hearts.

F. HYATT SMITH.

A GOOD PRAYER MEETING.

BY A. LAVERNE SPAFFORD, *Field Secretary Michigan Christian Endeavor Union.*

A good prayer meeting is a result of certain essentials that must be kept in mind if it is to be a success. Too many times, I fear, our prayer meetings are simply "get-together places" instead of having been planned to appeal to the highest and best there is in our beings. It is a training-school, but it is more than that. Not a meeting should close without having drawn some one closer to the Christ, or influencing lives to higher and nobler things. It should be a soul-winning institution, a stepping-stone to deeper consecration, for if the world is ever to be brought back to Christ it must be through such influences as are exerted through our prayer meetings.

With these few suggestions in mind, let us consider for a moment, three essentials, all of them magnified by our pledge.

First, if we are to have a successful prayer meeting, it will be because the leader and others have prayed earnestly about it for days and weeks. Endeavorers, you have no more important task or privilege in a whole year than leading a prayer meeting, and in your own strength you are not big enough. I believe in helps, but too many times we depend almost altogether on "printed helps" and forget from whence cometh our strength. "Prayer is the lever that commands the power of God." Let us use it. Pray not only for your own meeting, but for every leader and for every meeting as they are being planned. Do not let a meeting pass without observing the pre-prayer service. God is a prayer-hearing and a prayer-answering God, and will strengthen us for our tasks if we only ask. Prayer, then, is the first essential.

If a meeting is to be a success, it must be something more than the usual two songs, three-sentence prayers, etc. So, then, God will only help us as we plan. Planning is the second essential. If the prayer-meeting committee fails to suggest a meeting with the leader, then he should invite the committee to such a meeting. (In the live, well-organized society the latter will not be necessary.) Several minds are better than one, and the prayer-meeting committee should be ready to serve the leaders in helping to plan. Use *The Christian Endeavor World,* your denominational paper, and every good help available. Get all the help and information you can. With this fund of information stored away in your mind, lay the helps all aside as far as programme is concerned and put your own personality into the meeting. Make it your own meeting. Use clippings where necessary, but do not make your meeting a clipping meeting. Use them only with the younger, inexperienced members. Do not continue kindergarten methods endlessly with Endeavorers who ought to be in the high school. Advertise your meetings. Develop the talents of some Endeavorer who has ability in lettering or painting signs. If you have no bulletin-board for announcement, use a blackboard on the outside of the church, if possible on the lawn in a conspicuous place. If you have no lawn, ***make one.*** Have a good announcement ready for the Bible school each Sunday. Have it made by an Endeavorer who can be heard and understood. Make it snappy. Invite classes to come. Numbers increase interest and enthusiasm. Have your meeting planned so that there will be no dull moments. Have "minute-men" who will speak in the early stages of the meeting only when positively necessary.

In the third place, every meeting should have a *purpose.* No two meetings should be planned alike, because no two subjects are the same and are given for some definite purpose. Let us find what that purpose is and then plan to that end. If it is a missionary meeting, the meeting should not close without offering an opportunity for more definite service on the part of the society, or for some member to state his intentions as a Life Work Recruit. If it is a consecration meeting, it should stand for far more than a short devotional service and a mere mechanical calling of the roll. Perhaps there is some young person in the room who will take a stand as a Christian for the first time, or a member who will consecrate his life to definite Christian service. Do not miss an opportunity to clinch the results of the meeting. A purposeful meeting is as necessary as singing.

Pray earnestly, plan faithfully, and have a very definite purpose, and your meetings will show results that will be worthy of the name Christian Endeavor.

HOW TO HAVE A SUCCESSFUL SOCIETY—GOOD COMMITTEE WORK.

BY MISS MURIEL MCFEE, *President Rhode Island Christian Endeavor Union.*

It is impossible to rate too highly the importance of good committee work. Without it there can be no successful society. Again, through service our personal relation with God is enriched. And lastly, many an Endeavorer has been led into whole-time service for Christ and the church through definite committee training.

Successful work begins with the selection of the members by the nominating committee, a choice made carefully and prayerfully. Meeting with these folks thus chosen, we shall find, as often as the pressure of other duties permits, the pastor of the church and the president of the society, whose right it is to be notified of all meetings and whose privilege it is to act as counsellor and general helper.

Three essentials are to be considered for good committee work. First, know your job. Make it your business to know the general duties of the committee on which you have been placed (Efficiency leaflets and Expert

Endeavor). Add to this a knowledge of those specific duties which will place your society in the front rank of efficiency. Recognize the weakness of your society, that by so doing you may make that weakness its strength. Know your church problems, that you may by your committee work prepare yourself to fit in more closely with our church programme. Know what helps are available for our special department. From your conventions and rallies take away all the practical plans you can use. Know your job.

Second, plan your work. You can't drift into the best kind of committee work. Too many of our folks neglect to take a long look ahead, to make a general plan of the work to be accomplished. Then, carrying out first the plans immediately essential to the efficiency of the society, we can go on, one step at a time, until we accomplish the results we have desired. Let me urge you to have your committees organized, definite tasks for each individual. If you don't like your job, Mr. Committeeman, keep on with the game nevertheless. And you, Mr. Chairman, criticize kindly and praise generously the work that is being done.

Third, work your plan. Now we are ready to get under the load, and our efforts must be definite, systematic, not a hurry-up, last-minute matter, simply to provide work which may be reported at a too-near meeting of the executive committee. As we work, let us analyze what we have done, learn from our mistakes. We cannot afford to be unwilling to try out new methods; it is a part of the task before us. The average Christian Endeavor society has plenty of starters, and now we make a plea for those who in their committee work will stay through.

In every part of our committee duties—knowing, planning, working—comes prayer. A great electric sign flashed its message to thousands of people, "Prayer releases power." To us comes the definite knowledge that prayer does release power, and that without prayer we can do nothing. In our work we must constantly seek the guidance of Him whose work we do. So may we go on in His strength for the good committee work that makes for the successful society.

JAPAN FOR JESUS CHRIST.

By Rev. James H. Pettee, D.D., Tokyo, Japan.

1. I speak first in behalf of nearly 300 Endeavor societies with 6,000 members doing a quiet but exceedingly important service in the Far East, helping to hold the forces of righteousness working there loyal to the fundamental truths of our grand old gospel. And that is a tremendous task. When I think of the honored ex-presidents of our Japan Christian Endeavor union, Dr. Harada of Kyoto, for even years the able head of Doshisha University; Naomi Tamura, the fearless Presbyterian preacher and trenchant writer; Tokiyuki Osada, the St. John of the non-liturgical ministry of Japan, whose smile and hand-grasp and word of loving counsel have led hundreds of strong men to decide for Jesus Christ, I thank God that organized Christian Endeavor came to Japan twenty-seven years ago, and has continued its beneficent service there ever since.

Add to these the present official leaders, that broad-minded Methodist secretary, Mr. Ishizaka, one of the happiest combinations of strength and gentleness to be found in any land, who is acting president of the Japan union this year; and also of Secretary Sawaya, who is killing himself with overwork and struggling to make two ends meet on an allowance far too small for his own personal and family needs in these trying times of H. C. L. Why, he writes me that a Christian Endeavor tour which five years ago he could make easily for twenty *yen* now costs one hundred, and that the salaries of a host of government officials, including teachers, have been raised fifty per cent. during the past year. Time fails me to call the roll of Christian laymen of prominence in Japan, past or present, who are either the result (at least in part) of Christian Endeavor work, or have been associated with it in close co-operation.

2. But wider even than this circle and those whom they particularly represent, I speak to-day for 300,000 Christians in the Japanese empire. They know, none better, that Japan with all her wonderful achievements in the long past and her marvelous advances in these later years, Japan urgently needs the gospel of Jesus Christ. Without it she may easily become an angel of darkness and death to the Eastern world. With it there is nothing she may not accomplish as an angel of light and leading in the Orient.

3. Again widening my circle, I speak for the whole Japanese nation of approximately seventy-five millions of people, nearly sixty millions in Japan proper and fully fifteen millions in her dependencies.

The Japanese, a wonderful people indeed, who deserve the title of world leaders in their emphasis upon love of nature, physical cleanliness, simplicity of culture, an artistic temperament, courteous conduct, and loyal devotion to acknowledged leadership, are the Americans of the East in their alertness of mind and a desire to better their condition. They can mobilize their industries, their finances, and their philosophies as well as their armies with astonishing rapidity. They are the British of the East in their insular position, command of the sea and commercial ambitions; and they are the French of the East in their artistic skill, their punctilious politeness, and their tendency at times to value insular mannerisms more highly than international ethics. It must also in fairness be admitted that in recent years they have been the Germans of the East in advocacy of a centralized bureaucratic government and of militaristic expansion.

But Japan has come to realize as never before in her history that the sword, though the soul of the samurai, cannot be trusted to win the love of a people in subjection; that fine as are Confucian ethics, they do not cover the whole ground of modern needs; and that mystical and fascinating as are the doctrines of the Buddha, they are too overwhelming permeated with pessimism and fatalism to prove the inspirational guide of present-day spiritual leaders in world movements.

Japan has been honored, and rightly so, with a seat among the mighty as Versailles. She was the little finger on the hand that wrote the compact ending the world war. As such she represented much of Asia and added a touch of color to a white man's congress. It is regrettable that she did not win out on her main contention in her plea against racial discrimination, but alas! Europe, America, and especially South Africa were not yet sufficiently Christianized to desire or permit it.

The thing to be feared to-day is not that Japan will never redeem her pledge made to China in May, 1915, to return Shantung province to her big neighbor, but that western and so-called Christian nations will continue to discriminate so markedly against all Asiatic that the one Far Eastern people who have shown themselves ready and able to put up a strong and progressive government and a general national development along modern lines of activity will be embittered against her old allies and the ethical principles which they proclaim, and actually become a hindrance to the peace of the East, instead of a preserver thereof, which is her rightful rôle and her highest ambition.

We Endeavorers are asking, and with good reason, that the annual grant-in-aid to Japan from the World's Christian Endeavor Union be trebled from next year. If that be done, advance steps could be taken that would multiply many-fold the efficiency of our service in behalf of the churches and the children of the Sunrise Kingdom.

And one other consideration should be pressed home to the consciousness of every friend of the cause. In dealing with the Japanese we are working not for a savage or backward people, but with one of the dominant and most progressive races and nations of the day. Whether we like them or not, whether we approve of all that they do, is not the question before us. They have their grave faults as well as striking virtues.

The thing to remember is, they are a compact, homogeneous, well-governed, ambitious, powerful people. They are in a position of unquestioned prominence and tremendous possibilities. They urgently need the gospel of Jesus Christ. They half recognize that need. For the Christian church to fail them in an hour like the present would be a calamity indeed. It would go far toward being an ignominious confession that there is truth in the calumny, "Your religion can lift the lowest of the heathen and rescue the dregs of society, but it cannot save the highly refined or meet the exacting demands of an advanced civilization."

Not one of us believes this. Let us to the task, then, of disproving the slander by saving earth's best for higher service yet in the Lord Christ's Kingdom. God bless and save and sanctify the empire of Japan and make her truly great in sacrificial service, international fellowship, and spiritual as well as material achievements.

Tokyo, Japan.

THE JUNIORS: TRAINING FOR SERVICE.

By Miss Grace F. Hooper.

During the last two years many new words have been coined, and old words given new and fuller meaning. Two that have been on our lips more often than many others have been "training" and "service." We learned the meaning of "preparedness" in more ways than one. If the United States, that was to bring the higher and finer national ideals to Europe, needed trained leaders for her conquest, how much more does the church, bringing the gospel of peace and good will, carrying light to the darkness, and training not only for time but for eternity, need more fully equipped and better-trained leaders, that they may render the greatest possible service.

Did you ever stop to consider why men twenty-one to thirty-one were first called to the training-camps? Because they would respond more quickly and fully to the training than those older or younger. The greatest development in the shortest length of time could be realized.

Just as Uncle Sam was laying the foundation for the greatest service for the future, so must the church of to-day. There is just one place to train for future church leadership, and that is with boys and girls. There is just one organization in the church that makes a specialty of training leaders, and that is Christian Endeavor. Don't misunderstand me. I don't mean that it is the only worth-while organization, for I am not forgetting the wonderful work of the Sunday school; but its object is not to train for leadership. It is the impressional side of the child's development, while the Junior Endeavor is the expressional side, and all educators know that no character is a well-rounded one without both fully developed.

All churches of all denominations have wonderful new programmes before them, and the great call now is for leadership to carry them out. If through the past ages the church could have had the vision of its real mission in the world, that of sending out leaders, it might have been ready for the wonderful programme of reconstruction, but it was simply playing at the most important business in the world. We of this generation can do nothing with the church of the past, and very little with the church leadership of the present, but we have everything to do with the church of the future, for its leaders are in our midst now.

You can make out of boys and girls anything you want to, if you just want to "hard enough." But we have been waiting till the lookout committee of the Christian Endeavor went after them in middle and later 'teen ages, or until an evangelist came to bring them from ways of sin down the "Sawdust Trail." It's a splendid thing to reform, but a thousand times finer to form; then there are no scars, no forgetting, no harvesting of wild oats.

Training for *service!* Where? In devastated Europe. Wouldn't it be fine if we could send all the leaders needed from the churches and not have the vacancies felt at home? We are sending some fine ministers, Y. M. C. A., Y. W. C. A., and Christian Endeavor workers "over there," but for each one sent it means work left undone "over here." Wouldn't it be splendid if we had "enough to go around"?

Training for *service!* Where? In the foreign missionary field! Never were doors thrown so wide open and barriers brought so low and hands so stretched to America as to-day. Never was the call so clear, "Come over into Macedonia and help us." A call to one denomination came for the month of January from the one country of Turkey for one hundred and seventy-five new missionaries,—but the call couldn't be answered, for there weren't "enough to go around." This very day, if six hundred missionaries could land on the shores of China, there would be a place for every one of them; but there aren't "enough to go around."

Add to that the call from Japan, Korea, India, Africa, South America, Central America, and the Islands of the Sea, and you have something of the call for service from the foreign field.

Training for *service!* Where? In the home missionary field! We have counties and counties in the United States with but one minister to a county, and some with no minister to a county. We have pastors driving seventy-five miles a Sunday, and preaching three times, no place able to have two services a month, simply because there aren't "enough to go around." Schools for colored immigrants, mountaineers, Chinese, Japanese, Indians, poorly taught and poorly equipped, simply because there aren't enough men and money "to go around."

Training for *service!* Where? At home. In our own churches and Christian Endeavor societies. We hear the call on every side there for leadership. More and better-trained Sunday school teachers, members of woman's missionary societies, church official leaders, for our new reconstruction programme, officers and chairmen of our Christian Endeavor societies, heads of Quiet Hour, Tenth Legion, citizenship, missionary departments, Intermediate and Junior superintendents. Oh, it's a call for leadership all along the line—trained leadership.

Trained leadership for service. Do you need it? Do you want it? Would you like to have new members come into your Christian Endeavor "knowing the job"? Would you like to have new members ready for service rather than ready for training? Would you like to have leaders for your special departments, and have leaders to send out later on trained for the world service? You may have just that if you are willing to have a part in the training. As I said in the beginning, there is just one place to train for leadership, and just one organization in the church to do it.

You have but one life to live. Do you want it to count for the very most? Then invest it where the returns will be greatest. You can live in but one place and do but one work of yourself, but by helping train boys and girls for service you can put a part of your life in China, a part in the mountains of Kentucky, part in reconstructed France, part in your own Christian Endeavor society and church for years to come, and through these boys and girls trained for service your life can go on down the ages "richer, fuller, sweeter."

WHAT THE COMMUNITY HAS A RIGHT TO EXPECT OF THE CHURCH.

By J. T. McCrory, D.D., Pittsburgh, Pa.

This subject was assigned to me for discussion. It suggests an inquiry of great importance and interest—what the community has a right to expect of the church. The statement of the subject concedes the obligation of the church to the community. Well, what is the church, that it is debtor to the community? The church, to employ an apostolic phrase, is the body of Christ. As Paul puts it—and he was once thought to be good authority on a subject of this kind—as Paul puts it, those who become believers in Jesus Christ are by the very act of believing incorporated into the life of their Lord so that they are said to be *one* with Him. He is represented as the head, the life, the energy, while believers, the church, form the body through which He makes His life and power manifest. He Himself employed the figure of the vine and its branches to set forth the idea. The vine in which the life resides puts forth its energy through the branches. The fruit is found in the branches, but the vine is the life, the energy that causes the branch to blossom and bear fruit. The church, Paul says, is the body of Christ, and we are members in particular. To get at our investigation, then, we must know what the Lord proposed to effect through, by means of, His body—His church.

One of the very first statements He made concerning the matter is found in the report of His first formal discourse to His disciples in His Sermon on the Mount. He said there of the disciples: "Ye are the salt of the earth." "Ye are the light of the world." Now sin is festering, corrupting, putrescent, destroying, while salt is preserving, purifying, sweetening, restoring, saving. Sin is darkness, delusion, falsehood, moral, spiritual, intellectual, social blindness, while light illumines, discovers, informs, gives vision. The church, then, made up of people who are salt and light, is to save society. This is important and must not be overlooked—that, according to the word of our Lord, His mission, and the mission of those who represent Him, is to *save* the *world*—not some individuals, some souls in the world, but the *world*. "Ye are the light of the *world*." "God so loved the *world*." "God sent not His Son into the world to condemn the world, but that the *world* through Him might be saved." "Go ye into all the world and make disciples of all the nations." This mission of the church is a tremendous enterprise—to enlighten disciples to save the whole *world*. It means more than saving the individuals of humanity. It includes every legitimate enterprise in the world,—religion, business, education, politics, industrial interests, social affairs, government: it means, in short, the actual enthronement of Jesus Christ over all human affairs; it means the setting up of the kingdom of God in the world, the kingdom of righteousness and peace and joy in the Holy Ghost. Stupendous, glorious, divine. Now what has the community a right to expect of such a body charged with such an enterprise? It has a right to expect of the church that it make and fulfil its divine appointed end in the plan of its Head, and make good His claim for it as salt and light. To be more specific:

1. If the church is here to save the world, then it must exert regenerating power, for every other kind of salvation waits on the salvation of the soul-regeneration. To make good its claims to do what the community has a right to expect of it, it must exert converting power, for nothing is accomplished in the business of saving the world until the heart of the world is regenerated and the sinner converted. The new birth is not only necessary that the individual may see the kingdom of heaven, but also that society may be transformed into the kingdom of God.

We are very greatly interested to-day in the reconstruction of the world. The greatest statemen turn aside from every other great enterprise to take account of this supreme problem. Well, this is clearly the plan of God for the reconstruction of the world, this of the regeneration of the people. Ezekiel, speaking to the Babylon captives as to the plan and purpose of Jehovah under the new covenant put especial emphasis on the fact that it would be accomplished by the process of regeneration: "A new heart will I give you and a new spirit will I put within you. And I will put my Spirit within you." That is God's plan for the new world order. He is going to rule from within. He will write His law on the heart and put it in the mind. The kingdom of God in the world, the saved world, in other words, is to be a society ruled from within by the very Spirit of God.

Brethren, the hope of the world is Jesus. This is just as true of the social, industrial, and political order as it is of the salvation of the individual soul. When we come to consult the word of God, we find that the same one who was foretold as coming to open blind eyes, unstop deaf ears, make the lame leap as the hart and the tongue of the dumb vocal with song, is the same one through whom the weary centuries of war and political confusion are to be transformed into an era of universal brotherhood; tyranny, hatred, and oppression are to be banished from the families of the nations; and men are to learn war no more.

These splendid predictions in the prophets and the psalms that have inspired the hopes of men for better things through all the centuries, are but the forechant of a glorious anthem that is yet to be heard around this sorrowful planet. Let me tell you, the poet was not dreaming when he cried:

"O scenes surpassing fable, one song employs all nations,
And all cry, 'Worthy the Lamb, for He was slain for us.'
The dwellers in the vales and on the rocks shout to each other,
And mountain-tops from distant mountains catch the flying joy,
While nation after nation, taught the strain,
Earth rolls the rapturous hosanna's round."

Now if the church is to fulfil its obligation to the community, it must exert the power that will fit men for this high destiny. It must be able to take wicked men, selfish men, stony-hearted men, evil-doing men, and make of them good men, tender-hearted men, God-fearing men, God-obeying men, in all the purposes of God for men in this world. And if the church in any community is not exerting that power, that divine, dynamic, regenerative power, it is not doing what the community has a right to expect of the church. It may be doing a great many other good things, worthy things, benevolent things; but if it is not doing this thing, it is not meeting the just claims of the community. And this is just as true of the church in all the world as it is of the church in any community. If the church in the United States is not putting forth converting power and getting men ready for the kingdom of God in the world according to the terms of the new covenant, then it is not doing the work for which it was ordained.

2. The community has a right to expect of the church that it promote universal good will and brotherhood among men. However we may read the song of the angels at the coming of Jesus, it means that He came to bring about good will among men as well as glory to God. In fact, the glory to God from the advent, so far as this world is concerned, depends largely on the measure of good will it produces among men. There never will be a true kingdom of God on this earth so long as men hate one another, antagonize one another, war against one another, in any of the relations of life. Now somebody must teach the way and set the example of true brotherhood and heavenly good will if the kingdom of God is ever to be a reality in this world; and who is there to do that but the church? And if the church is to do that it must show it in its life. If the people who profess to be controlled by the mind of Christ cannot live according to this rule, what right have they to ask others to follow them? In the degree in which we believers illustrate by our lives the spirit of Jesus Christ in our relations to one another and to all men, in that degree will our testimony in any community be influential. And when I speak of the church in a community I do not mean any one particular denomination of believers—I mean the church, the whole body of believers; all the people who are joined to the Lord and compose His body—the called-out company by whatever name they may be designated.

Now it is not in my heart to begin a tirade against sectarianism, and say that if the church is not meeting its claims in this regard to the community the blame is to be laid at the door of denominationalism, because

I do not believe that is true. The fact is that if the mind of Christ possessed His disciples, they could manifest it just as fully under different names and different forms of church order as they could if they were all merged into a single order. If I read history correctly, the time was when the whole body of professed believers were in one order and of one name, and that a mighty good and suggestive name, for they called themselves the Holy Catholic Church; but there never was a time when the angels' song was so much of a misnomer, when there was so little of real brotherhood, so little of the spirit of Jesus Christ, in the church.

I am not sure that we ever want to see Christendom again ruled by a single ecclesiastical organism, for when one did rule it was for centuries the nightmare of human governments and the grief and horror of all devout souls, and became the agent of a tyranny so monstrous and so unmitigated as to call down upon itself justly the maledictions of an outraged and oppressed humanity, the hatred and anathemas of a robbed and misruled world. No, I am not sure that we ever want to see one great ecclesiastical order for believers; anyway, I am pretty sure that we will wait a long time for it if it is to come. Blood is thicker than water, and when one recalls the rise of the great historic denominations, he is persuaded that it will be a long time before earnest disciples will be weaned from their attachment to their ecclesiastical ancestors and devotion to the names they bear.

Why, when I let my heart have its way and look back yonder to the morning of the glorious Reformation, and see Martin Luther standing up there single-handed and alone against an apostate church and a godless world, giving back at the risk of his own life to the world the great doctrine of justification by faith and putting to shame the courage of a dozen centuries by the splendid declaration, "Here I stand; I can do no other; so help me Almighty God," I almost wish I had been born a Dutchman instead of a Scotch-Irishman. But then I think again of my own ecclesiastical forbears, the dissenters and Covenanters, who bedewed all the heather and enriched all the soil of both Scotland and Ireland with the most precious patriotic blood that ever ran hot in patriot hearts or leaped living from patriot veins to consecrate a cause or maintain the "crown rights of King Jesus," then I am glad I belong to the great Presbyterian family. Then I am reminded of the church that came up through seas of blood and gave us the King James translation of the Bible, the church of Cranmer, Wilberforce, Gladstone, and Phillips Brooks, and I say, "From the man who can see nothing in the great Episcopal denomination to thank God for—from such a man, Good Lord, deliver us." Then I am reminded of the Puritans and the Pilgrim fathers, those men who

> "Could not live by man-made codes and creeds;
> Who sought the path where every footstep bleeds;
> Protesting, not rebelling; scorned and banned.
> Through pains and prisons, harried from the land"

coming to these wild shores with savage beasts and still more savage men, that they might found a church without a bishop and a state without a king, and I say, "All honor to the Congregationalists." But right among those New England Congregationalists I see Roger Williams protesting against the interference of the magistrate in matters of religion and insisting on a church of spiritual persons, and going farther off into the wilderness that he might enjoy spiritual liberty, and I say to my Baptist brethren, "You stand for a principle for which it was a glory to live and would be immortal fame to have died." Then I stretch out my hand across five generations of the past, and touch but the tip of the finger of a man who was so surcharged by the Holy Ghost that single-handed and alone he aroused a faithless, dormant church and startled a godless world, and organized under protest a new ecclesiastical fellowship which has been an inspiration to all Christendom and has sent millions sweeping through the gates of the New Jerusalem washed in the blood of the Lamb, and I say, "Who would not glory in the name and fame of Methodism?"

No, I am not sure about the advantage of one great, world-wide fellowship embracing all of Christendom, but I am sure that until the Spirit of the great Head of the church fills and inspires all its vast membership it will not meet the claims of a community or of the world to set the example and lead the way to perfect good will and brotherhood, and this it is under bonds to its Lord and to a world full of discord, dissention, and faction to do.

3. The community has a right to expect the church, with her heavenly as well as earthly ideals, to so lead in the great work of intellectual uplift and development as to ensure for education that it will be permeated by the very Spirit of Jesus Christ, and will thereby ensure the largest liberty of thought and investigation as will educate men to think God's thoughts after Him. That is the kind of thinking the world needs in its science, art, religion, literature, history, philosophy, industry, music, poetry. What the world needs and what the mind of man was created for is just to think the thoughts of the Creator after Him. God has written His message for man in creation and providence as well as in revelation, and He wants men to discover and glory in His thoughts. The heavens declare His glory, the sun and stars, the seas and tides, the rocks and rivers,—the whole vast realm of nature is the transcript of the mind of the Eternal; and education should be just an earnest, reverent effort to hear His voice, discover His mind, and glorify Him by following His wisdom in all the affairs that have to do with morals. But that never can be done in schools that trifle with His revelation, ignore His Son and His salvation. The church will never meet the needs of the community as salt and light so long as the dominant forces in education are indifferent, infidel, or unbelieving. Then there are forces directly attacking man-soul, crowding in through eyegate and eargate and every possible way of entrance that threaten its very destruction unless they are counteracted. Unless the church in any community is on the alert, it will utterly fail to meet its obligations as the salt and light of the world.

4. The community has a right to expect of the church that it stand for and promote the highest ideals of community life. That society must be reconstructed if the world is really to be saved is beyond argument. The confusion in the social order, the selfish rivalries, bitter contentions, warring factions, and all that must give place to love and good will in the salvation of the world. Take the matter of capital and labor, or employer and employee, or however you want to think of this endless conflict between those who labor with their hands and those who provide and control the opportunity for labor. Why, the contention that results, if let run its course under the dictates of selfish human nature, would issue speedily in Bolshevism or anarchy, in the utter destruction of all social order. But the church announces to the distracted world that under its direction the social order is headed toward the millennium. She holds out the blessed promise to all men of a state of society in which the abundance of the divine bounty shall be so distributed that every one shall have enough and to spare, and all shall be satisfied with the distribution. Now if any one contemplating society as it is to-day should be disposed to laugh at such a proposition as idealism, all I have to reply is that it is idealism, Christian idealism; but if that is not what the gospel promises to bring about in this world, then I am frank to confess that I do not know what it does promise to bring about. But if that is its promise, who, then, is to set the example and lead society toward that glorious consummation but the church of Jesus Christ? Then I must insist that any community has a right to expect the church, the Christians of the community, to set the example and lead toward the ideal condition. If some one protests that it is expecting too much of poor fallen human nature, then may I ask what we are to expect? Is it not out of this poor fallen human na-

ture, regenerated and renewed after the image of the Christ, that the saved world is to be constructed? Brethren, I confess to you that the thing appears like an utter absurdity from the human standpoint, and it is only when I get my courage screwed up to the sticking-point where I can believe God that I dare preach this gospel of world salvation. Now what shall we do about it? We must either take our stand with Jesus Christ and insist on His church, His body, undertaking all He has promised for the world, or give it up and turn the job over to somebody who will believe Him and trust Him for the grace and power to undertake the impossible. I assume that Jesus knew the difficulties in the way when He undertook the salvation of the world; but He undertook it and laid the responsibility for its accomplishment on His body, His church, and guaranteed all the resources necessary to complete the job, and by the grace of God I am going to trust Him until the heavens fall or you can prove to me that the pillared firmament is built on stubble. The community has a right under the gospel promise to expect of the church that she set the example of, and lead the way to, ideal social conditions.

5. There is just one more thing I should like to suggest, and that is that the community has a right to expect that the children of the community shall have the especial attention and nurture of this heavenly society. Nothing for a community can surpass in importance the care and nurture of its children. When Jesus said, "Suffer the little children to come unto me and forbid them not, for of such is the kingdom of heaven," He meant to challenge the attention and interest, not alone of that little group of disciples, to the place of children in the extension of His kingdom, but of all disciples everywhere and in every generation. Statistics warn us that vast numbers even of the children of the church are lost to the kingdom because of unchristian conditions in Christian homes, while the community has a right to expect of the church that she exert a saving influence on those beyond her pale. But if we are not able to save our own children, how are we to save the world? This is the age of the children. Never before were children receiving the attention they are receiving to-day. Surely the church should charge itself with the high eternal interest of the young of the community to such a degree that there would be thrown about them the most wholesome and elevating influences. The home life of the members of the church should be such as would meet the high claims of the founder of the family for the home.

But I have said enough. Already methinks I hear good, earnest Christian people protesting that we are expecting too much of the church. We are told that at the very best we are less than one in four in any community, and it can hardly be expected that we should be held responsible for the impossible. But that is just what Jesus has us in the world for—to accomplish the impossible. The little band of a hundred and twenty at Pentecost faced a whole hostile world, but Jesus sent them out to the work of saving the world. Samson was only one man in a million, but as long as he kept in touch with his source of power he was able to accomplish the impossible. It was only when he laid his consecrated head in the lap of luxury, alias Delilah, that he lost his vision and his power to do the impossible. Gideon was acting like a coward when the angel found him out behind the wine-press with his flail and his newly-gathered sheaves of harvest for fear of the Midianites; but when the Almighty took him in hand he developed such power and courage that the fame of his exploit is still ringing through the centuries. God knows we are weak, a mighty sight weaker than we think we are. But it is when we are weak that we are strong. It is when we come face to face with our impossible task and the claim of a dying, doomed world for the exercise on its behalf of power we do not possess that we are driven to the source of power, all power, in the person of our enthroned and omnipotent Lord. "Ye are the salt of the earth." "Ye are the light of the world." Shall we shrink from the impossible task assigned us of enlightening and saving the world?

Dr. Daniel A. Poling,
Speaking at the Canadian Rally.

CHRISTIAN ENDEAVOR WEEK.

By Rev. Robert P. Anderson, *Editorial Secretary of the United Society of Christian Endeavor.*

I do not need to attempt to prove to an audience like this the advantage of celebrating Christian Endeavor Week. We believe in celebrations, just as our forefathers did before us. The Jews had their great commemorative festivals, the Feast of Tabernacles, which lasted a full week, the Passover, Purim, in memory of the deliverance of the nation under Esther, and others. The Pilgrim Fathers celebrated Thanksgiving, and to-day we have Independence Day, and Memorial Day, and Easter, and Christmas, and New Year's, and Mothers' Day, and unless signs fail we shall soon have Fathers' Day also.

Our Christian Endeavor celebration is not merely commemorative. Our faces in this time of all times are turned, not to the past, but to the future. Had Christian Endeavor needed a commemorative festival, one day might have been enough for it. But one day is not enough in which to compass the future. A week is short enough in which to emphasize a world programme. With the growth of Christian Endeavor itself its celebration grew from a day to a week.

It was at Los Angeles in 1913 that a week's celebration was decided upon, and the dates were so fixed as to include the Sunday before and the Sunday after February 2, the real birthday of Christian Endeavor.

The programme which the United Society proposes for the Christian Endeavor Week celebration in 1920 differs a little from the programme of former years. Up to this year we have fixed the character of the various days in the week, giving a certain day to unions, another definite day to church loyalty, and so on throughout the week. This time, that is, for 1920, we propose to fix only two days in the week, Sunday, the opening day, Christian Endeavor Day, and Sunday, the closing day, Decision Day. The value of these two days has proved very great, and the climacteric services of Decision Day in particular have been widely blessed.

Besides these two days we outline suggestions for six more days, but do not attempt to fix the days. The local unions and societies will make out programmes of the week's activities, arranging the days to suit local conditions. Thus Fellowship Day may be celebrated on any day of the week which the local union determines upon. Church Loyalty Day will fall on the day of the church prayer meeting. Social Day will be celebrated on the day which the unions or societies consider best for social activities, and that day will vary in different sections of the country and in different cities. The list of days suggested is as follows:

1. Sunday, Christian Endeavor Day. 2. Sunday, at the close of the week, Decision Day. Then the following, the order to be changed at will:

3. Fellowship Day, for union rallies, for Alumni celebrations, for reunions, and for the consideration of vital topics of the day such as temperance, world peace, the Interchurch World Movement, and so on.

4. Junior Day, for which Saturday is suggested, with a Junior rally, a model meeting, and a demonstration to parents of the value of Junior training.

5. Intermediate Day, which may be celebrated Saturday, or Sunday afternoon, or any day in the week. This day should include a banquet, a snappy rally, and a social hour.

6. Social Day, for a social to which officers of church and Sunday school as well as former Endeavorers should be invited.

7. Extension Day, which should be the culmination of weeks and even months of preparation for the winning of new members and the organization of new societies, Junior, Intermediate, and Young People's.

8. Church Loyalty Day, to be celebrated on the day of the church prayer meeting, the Endeavorers being given a definite part in the programme of the meeting and working up the attendance.

The United Society expects to issue in pamphlet form programmes for several of these days. The pamphlet will contain a programme for Fellowship Day, Junior Day, and Intermediate Day, besides programmes for Christian Endeavor Day and Decision Day.

To harvest the blessing in Christian Endeavor Week societies must begin their preparation for the celebration early, at least a month in advance of the date. Christian Endeavor Week will be just what we make it. If we get the programme pamphlets and plan the week in advance and pray over and push through the plans that we make, the week will bring inspiration and power.

This next celebration should be the best ever held. The boys will be back from the front, most of them. The meetings should be filled with thanksgiving, with rejoicing, with hope. We stand at the gateway that leads to a new world. A crisis is upon the world and a challenge has come to the young battalions of Christian Endeavor. How shall we meet that challenge? Shall we respond with high courage and let our Christian Endeavor Week's celebration ring with the note of triumph?

This year of all years Christian Endeavor Week should be Victory Week for the young people of our land.

CHAPTER XVIII.

A GROUP OF REPORTS.

REPORT OF DANIEL A. POLING, LL.D.

Associate President and Citizenship Superintendent of the United Society of Christian Endeavor at the International Conference Buffalo, N. Y., Wednesday, August 6, 1919.

To the Board of Trustees of the United Society of Christian Endeavor:

Gentlemen:

Since January, 1918, I have spent fourteen of the nineteen months in the field, seven months of this time in connection with two overseas journeys. This report will not deal with general observations or with recommendations, but will confine itself to Christian Endeavor as I found it in war service, and to the growth of the Alumni Association. The statement of the citizenship department is included in this report.

In England, France, and Germany, I was associated with representatives of more than two hundred thousand Christian Endeavorers who rendered our country military service, and in all of these countries I met our brothers and our sisters in the uniform of the Y. M. C. A. and the Red Cross. Numbers of them paid the price of victory with their lives. Others bear upon their bodies burning wounds. The highest honors of war rest upon the breasts of many. The first Croix de Guerre awarded to an American after we entered the conflict was given to a Christian Endeavorer, Robert Patterson, of Huntingdon, Ind., who enlisted as a private and rose to a lieutenant-colonelcy in the artillery. Justin P. Follette, president of the San Diego Christian Endeavor union is one of those who received the Distinguished Service Cross.

From the meagre executive force of the United Society (eight) three gave a total of twenty-three months exclusively to war service, two of these overseas and one in home training-camps.

Christian Endeavor throughout the war, true to its policy and tradition, functioned through authorized and accredited agencies of the church. In the interests of economy as well as of efficiency, it invested its money and its services through the Red Cross and the Y. M. C. A. It did not allow pride of organization to persuade it into a selfish, spectacular, organizational programme. The men and women furnished by it to the institutions overseas authorized by the Government and recognized by the churches were its pride and its glory.

It will not soon be forgotten that the United Society of Christian Endeavor was the first religious organization to pledge its support to the Government when war became inevitable. It is a source of gratification to the officers of the organization that this pledge was made without hatred against any people.

All that has been written here is related to the activities that are naturally classified under the citizenship department. The United Society superintendent of this department was chairman of the committee named by the Federal Council of Churches to confer with the War Department and if possible to promote a programme of temperance education for the army and navy. This committee, after several months of labor, was successful in organizing the United Committee on War Temperance Activities in the Army and Navy, which throughout the war represented all national temperance groups, Catholic as well as Protestant. It was directly under the sanction and direction of the War Department as a subcommission of the Fosdick Commission, with leaflets, posters, moving pictures, animated cartoons, stereopticons, stereoscopes, stereomotorgraphs, special electrical signs and devices, and with addresses by many of America's most famous publicists. The committee kept before the enlisted men on both sides of the Atlantic the claims of decency, morality, and temperance.

The plan was wise in this, that the programme of the committee was not allowed to stand alone, to face the prejudice the programme of a temperance organization as such would have been compelled to meet. It came everywhere to the army and navy with the challenge of patriotism and as a part of the general, patriotic, religious, athletic, and moral activities of the service.

More than $200,000 was received by the committee, with virtually no appeal to the public. The secretarial services which it commanded from the various welfare organizations and from the military represent in actual value considerably more than a million dollars.

The work is now practically completed. A small balance remains in the treasury, and a record of service has been achieved that time will reveal. Certainly millions of men have been prepared for national prohibition, and have also learned the reasons for the nation's purpose to keep her fighters physically competent and morally fit.

In addition to furnishing this committee, which was the only organization allowed to carry on temperance educational work in the army and navy, with its chairman Christian Endeavor supplied it, with its two executive secretaries, Rev. Harley Gill, of California, and Mr. Arthur Whitney, of Des Moines. When it became necessary to open the Paris office, Mr. Whitney was made overseas secretary.

The citizenship superintendent of the United Society spent four weeks in a speaking campaign in England and Scotland under a British educational campaign. He brought back from Great Britain a personal letter from Premier David Lloyd George that is the only direct comment thus far made officially by the British government on America's prohibition activities. This letter directly influenced the attitude of the American Congress on the intimated embargo on foodstuffs to Europe proposed as a result of the continued use of grains in the overseas manufacture of intoxicating liquor.

The United Society of Christian Endeavor is still honored with the presidency of the National Temperance Council, and let us thank God that the resolution written by Drs. Grose and Landrith, and adopted by the Atlantic City International Convention of 1911, calling for "a saloonless nation by the Fourth of July, 1920, the three hundredth year from the landing of the Pilgrims at Plymouth," has not only become the slogan of all temperance societies, but has become a fact of government. "A Salooonless World by 1930," sent forth from Chicago in 1915, is to-day the battle-cry of prohibitionists everywhere.

The future of our citizenship department must be filled with campaigns of Americanization; while we continue to strike out against the evils that menace youth and corrupt politics, we will not forget those more vital constructive programmes that have as their objectives the building of a citizenship fit to govern and wise to rule.

Since January, 1918, I have travelled 75,690 miles. Of this total, 32,090 have been covered since the beginning of 1919. Since my return to America in March I have devoted my time, aside from visiting eight State Christian Endeavor conventions, to the promoting of the Alumni Association. In the Alumni tour of 91 days, 20,090 miles were covered in 26 States; 164 addresses were delivered in 66 cities to an aggregate attendance of 78,700. Fifty-three Alumni fellowships were organized, with a total annual plédge for the world programme of the United Society of Christian Endeavor of a little more than $13,500.

There are now 84 local Alumni fellowships, with a total annual pledge for the work of the United Society at home and abroad of $20,000, of which amount $5,000 is returned by the United Society to the various States, to be used exclusively for State field-work. These Alumni Associations are scattered from Massachusetts to California, and from Texas to Canada. We have just begun to realize the possibilities of the plan. A mine has been opened that under God, and by careful working, will provide support for the modest but imperative programme for which Christian Endeavor is now and will be increasingly responsible.

It would be presumptuous indeed for me to bring a statement in this presence of the field, the mission, the genius, of Christian Endeavor. You have lived with the movement since it began to be. But for myself in the past eighteen months I have come to accept with new and larger faith its essential, its unique place in the plan of God for His kingdom. The executive officers believe that the Alumni will make possible the larger usefulness of our movement in which they have been blessed, and whose blessings they now pass on to those coming after them. Without appeal to local societies, with only a modest appeal to a comparatively few individuals who enrolled as United Society Alumni, we shall be able to finance our activities on both sides of the two seas.

Particularly gratifying has been the support of denominational and State leaders. That this programme will materially add to the Christian Endeavor contribution to denominational activities has already been demonstrated. Of this, Secretary McLean has written as follows:

"What is given for the support of these workers is not so much taken from the missionary treasuries of the missionary societies; rather, what is given should lead to an increase in the gifts of God's people for every worthy cause."

But of more importance than financial support is moral support. The Alumni Association is building up in scores of great communities a body of Christian Endeavor loyalty among the older men and women that is a strong stay for the active union leaders and a bulwark for the movement at large.

The original Alumni Association, Fellowship No. 1, of Cleveland, which has honored the writer with its support, and which made possible his two overseas trips, has now a membership of more than 500, with an annual pledge of approximately $3,500. This in less than four years. At its annual banquet in April of this year 339 sat down together. Some of the city's most distinguished personages were present, and an appeal for enlarged financial support was made by the moderator of the Cleveland Presbytery.

"The Lord hath done great things for us, whereof we are glad."

TWO YEARS OF PROGRESS.

REPORT OF WILLIAM SHAW, LL.D.

General Secretary of the United Society of Christian Endeavor, Given at Buffalo, N. Y., Wednesday, August 6, 1919.

The two years' reviewed in this report cover practically the period in which the United States was engaged in the world war. It was a time of testing such as we had never before experienced in the history of our movement. In the United States alone more than two hundred thousand of our Christian Endeavor boys were in the service, and probably not fewer than one hundred thousand more in the ranks of the Allies. At least as many more of our Christian Endeavor girls were engaged in some form of active war-work that claimed a large part of their time and interest. It was a time of abnormal activity, when the claims of regular service had to give way to the imperative demands of the hour. In addition to this there was the disruption of our work in local societies and unions, caused by the coal famine and the influenza epidemic. In spite of these and many other adverse influences the years under review have been two of the greatest in point of achievement and progress that Christian Endeavor has ever experienced.

Eight years ago, at the Atlantic City Convention, Christian Endeavor issued the challenge, "A Saloonless Nation by 1920." The liquor men sneered at it as an example of youthful effervescence, and the Christian people questioned the wisdom of setting so impossible and impracticable a goal. But God was in it, and with Him nothing is impossible. Within two years "A Saloonless Nation by 1920" became the slogan of all the moral-reform and religious organizations of our country. It gave a definite goal toward which to work, and State after State wheeled into line for State-wide prohibition. The sentiment became so overwhelming that Congress yielded to it, and the constitutional prohibition amendment was adopted and sent back to the States for ratification. In about one year forty-five of our forty-eight States by great majorities ratified the amendment; and January 16, 1920, the United States goes "dry," never again to enter into partnership with the most iniquitous, lawless, treasonable traffic that ever cursed a free people. In the meantime war prohibition came in to hasten the dethronement of King Alcohol in common with all the other kings and kaisers who lived by exploiting the people. Let us work and pray that he may not be galvanized into life for a few months only to pass through the agonies of a second death.

July, 1918, saw the completion of another commanding Christian Endeavor enterprise when the World's Christian Endeavor Building on Beacon Hill, Boston, costing with the land nearly a quarter of a million dollars, was dedicated. Beautiful and useful, it stands as an expression of appreciation from nearly one hundred thousand Endeavorers of what Christian Endeavor has done for them. A power-house of spiritual forces, its influence radiates into all the world.

One of the most significant and far-reaching movements in our history is the organization and development of the Christian Endeavor Alumni Associations, to which Dr. Poling has given so much of his time and of which he will give a full account in his report. In my opinion the Alumni Association is the one successful solution of the problem of adequate support for our world-wide work.

Another interesting development is the inauguration of the union tithe. Christian Endeavor has done pioneer work in challenging the individual with the principles of Christian stewardship and the acceptance of the tithe as the minimum gift for God's work. We are now asking ourselves, "If this is good for the individual, why is it not equally desirable for the organization?" Already sixty-nine Christian Endeavor unions have accepted the principle, and are tithing their budgets for the extension of Christian Endeavor in mission lands through the World's Christian Endeavor Union.

The Tenth Legion has enrolled 4,900 new members, and now has a total enrollment of 52,753 members who are giving not less than one-tenth of their income for the extension of God's kingdom on the earth. Here is one reason among many for the increasing gifts of God's people for the extension of His kingdom.

The wonderful growth of the Tenth Legion is perhaps due to the fact that, like the disciples of old, "they first gave themselves." Wherever the principle of the Quiet Hour has been accepted, there the Tenth Legion has flourished. Christian Endeavor has always emphasized the fact that being precedes doing, that a deep spiritual experience is the first essential. This can be maintained only by constant communion and fellowship with Christ, the revealer of God. As a result 16,817 new Comrades of the Quiet Hour have been enrolled, and we now have a total enrollment of 173,523

Comrades who know by personal experience the blessedness of practising the presence of God.

Along with this emphasis upon spiritual realities there has been a corresponding development along practical lines of service. Five thousand two hundred and forty-seven Endeavorers have successfully passed the examination in "Expert Endeavor," and have received the degree of C. E. E. We now have 24,844 Christian Endeavor Experts. Many unions conduct Expert classes, and the Maryland Christian Endeavor union has taken an advance step by establishing a correspondence course in Expert Endeavor.

"What will you do with your life?" is another of the challenging quesitons that Christian Endeavor is putting up to our young people at the very time when life-decisions are being made. Four thousand eight hundred and sixty-eight have covenanted with God that they purpose so to shape their life-plans that they may give themselves to full-time service for Christ and the church. Some of these Life-Work Recruits are already in the ministry, Y. M. C. A. or Y. W. C. A. service, or on the mission fields of the world. Many are in college, training-school, or seminary, while others are preparing for such a course.

Christian Endeavor is ministering in the name of Christ in fresh-air camps, hospitals, prisons, and other institutions. Societies have beeen organized in army camps and cantonments, and even at the battle-front. Our ships of war have carried groups of Endeavorers who have been loyal to Jesus Christ in the face of all the discouragements and obstacles to be found under such conditions.

The Christian Endeavor Army of Universal Patriotic Service, a half-mission strong, did fine service during the war. Its war-garden and food-conservation campaign received highest praise from official headquarters in Washington. The Illinois Christian Endeavor union received a special banner from the Government for successful work in selling War-Savings Stamps.

In order to prevent duplication and waste most of the work was done in co-operation with the Red Cross, the Y. M. C. A., and the Y. W. C. A. "Huts" were furnished, entertainments provided, meetings conducted, and books, phonograph records, and general supplies contributed. Our associate president, Dr. Poling, and our extension secretary, Dr. Landrith, were among the most popular and helpful speakers in camps and cantonments.

Dr. Poling made two visits to the front in France, and was gassed while serving our boys in the front-line trench. Mr. Clarence C. Hamilton, our *Christian Endeavor World* field-representative, rendered value service as a Y. M. C. A. secretary in France, and many of our State leaders served in the same way.

The quality of our Christian Endeavor representatives is shown by the fact that the first Croix de Guerre given by the French government to an American officer after we entered the war was given to Lieutenant Robert C. Patterson, an Indiana Endeavorer, who entered the service as a private, and is now a lieutenant-colonel. The first Australian to receive the Croix de Guerre was also a Christian Endeavorer.

Successful campaigns have been conducted to strengthen the Sunday school and deepen the interest of our Endeavorers in world-wide missions. So great was the response along the latter line that the supply of text-books for the mission study classes were exhausted in the first two weeks of the campaign.

Our Standards Campaign was greatly interrupted by the abnormal war conditions. Some of our unions lost every young man on their executive committee; but the young women rallied magnificently, and splendidly carried on the work. We have had reports from twenty-four Superior and fifty-three Excellent unions; one hundred and ninety-seven Superior and four hundred and twenty-six Excellent Young People's and Intermediate societies; and ninety-five Superior and two hundred and thirteen Excellent Junior societies. Five thousand eight hundred and fifty new societies were organized, and we now have a total of 77,657 societies and 3,882,850 members in our world-wide fellowship.

This year we have had some of the largest and most successful State conventions ever held by Christian Endeavor. California had a paid registration of 5,083, and Ohio 4,071. Many other States had the largest attendance in their history.

Christian Endeavor has had a remarkable growth in the Southern States, of which Secretary Lehmann will give details in his report.

The Intermediate and Junior work has received a new impetus under the inspiring leadership of our national superintendents, Mr. Paul C. Brown and Miss Grace F. Hooper. During the past year Miss Hooper has given her whole time to Junior field-work, and Mr. Brown has conducted several effective campaigns for Intermediate Christian Endeavor.

The organization of the new Southwestern Christian Endeavor Federation, including Texas, Oklahoma, Colorado, Utah, Arizona, and New Mexico, under the leadership of W. Roy Breg, secretary, has in it great possibilities. Watch Christian Endeavor grow in the great Southwest, the land of boundless opportunity.

The perior under review closes with the addition to the headquarters force of Mr. E. P. Gates, formerly field-secretary of Illinois, who becomes field-secretary for the United Society, his work for the first year largely centering in Illinois and Indiana, with a campaign in Massachusetts; and Mr. Stanley B. Vandersall, general secretary for Ohio, becomes Alumni superintendent for the United Society.

What is true of the United States is equally true of foreign lands. Mexico and South America are stretching out eager hands for Christian Endeavor. We face a challenging opportunity in all parts of Latin America. England and Scotland have put secretaries in their home field, and have started campaigns for more than $50,000 adequately to finance the work. From the great mission fields of India, Africa, China, and Japan come most encouraging reports and requests for enlarged resources that the opportunities may be more adequately met.

In view of these achievements it is difficult to understand the efforts of some Sunday-school and religious-education leaders to eliminate the young people's society and graft its work on the Sunday school. To carry out a theory or develop a system, they would disrupt or destroy the young people's movement to which the Sunday school and the church are indebted for many of their most devoted and efficient leaders. We believe that there is a distinctive field for each of these organizations, and that success lies along the line of co-operation and not of competition or absorption.

Those who have been nearest to the fiery heart of the recent world-conflict predict that out of it is to come a new world,—new politically, new industrially, and new religiously.

They claim that it is to be democratic, brotherly, a world of service. How like that is to the words of Jesus: "One is your Master, even Christ; and all ye are brethren." "He that is greatest among you shall be your servant." "The Son of man came not to be ministered unto, but to minister."

Christian Endeavor will require no change to adapt it to the requirements of the new order. It is democratic in every fibre. It is run from the bottom up, not from the top down. Don't lose sight of that fact, for it is to be the spirit of the new day. There are some who would change its spirit and put it under autocratic direction. Such a course would sound its death knell.

It is international, interracial, interdenominational, brotherly. Denominational loyalty and interdenominational fellowship are two of its fundamental principles.

One of its earliest mottoes was, "Saved to Serve." It has stood for service as the test of faith. It was one of the pioneer religious organizations to emphasize social service, but it never made the mistake of substituting service for salvation. The larger your social-service programme is, the greater must be your emphasis upon personal salvation. Christian Endeavor challenges to individual service, and trains in team-work.

It is the training-school of the church, that she may be ready to meet the demands of the new day, the new order, the new world. Let us rejoice in the privilege and gird ourselves for the new tasks.

FINANCIAL STATEMENTS FOR YEAR ENDING MAY 31, 1919, PRESENTED AT THE ANNUAL MEETING OF THE UNITED SOCIETY OF CHRISTIAN ENDEAVOR,

BY A. J. SHARTLE, *Treasurer and Publication Manager*, Buffalo, N. Y., August 6, 1919.

If it is true that the average corporation existing for material success is surrounded by water representing stock, then the United Society of Christian Endeavor Corporation is in a separate and distinct class. It issues neither stocks nor bonds, declares no dividends on its business, and, if surrounded by water, it is due entirely to its demand since 1911 for a saloonless nation by 1920.

However, no corporation, whether in business or religion, can evade the ever-present and always necessary systematized, numerical facts of a financial report, if it would be informed. And yet possibly the most unpopular inanimate things that attend conventions and conferences are the columns of hard, cold, uncompromising figures. Stripped of all embellishment, figures are not conducive to an attractive reception; but we need them in our business, due to the fact that they represent money, a driving power behind an organization that strives to promote the lives of young people to a large usefulness in the service of God.

My business as treasurer and publication manager of the United Society is to receive, get, make, and hold on to money. Consequently, the regiment of figures representing the financial reports which you have in hand diagnose the condition of the United Society Corporation as conducted with profit and no loss, thereby eliminating failure and establishing success.

However, permit me to remind you that it is not so much an organization's money that commands respect but rather its ability and achievements. But that is another story for others to relate.

We have just closed a year fraught with conditions of war, consequent business depression, epidemics, closed churches, cancelled meetings, and misdirected transportation facilities. We have played the game of business and religion in the arena of business with a clear conception of the need for more business in religion, and the opportunity for business to embrace religion. And we are not unmindful of the fact that in passing through this period of world readjustment, and contention with a grade of business that savors of the profiteer, we can still do real business and remain real Christian. Therefore, the facts that follow represent a brief statement of a thing done, and existing, and of a result achieved.

As custodian of all assets amounting to nearly one-half a million dollars, the property of the United Society of Christian Endeavor Corporation, we would report as follows:

Receipts from all sources for the fiscal year ending May 31, 1919, $78,926.60. Expenses, $76,400.42. Balance, $2,526.18. The assets of the corporation as of May 31, 1919, are $434,193.23. Liabilities, including annuities, $216,854.59. Net assets June 1, 1919, $217,338.64. Ten years ago the assets of the United Society were $61,835.16. The large increase in assets is chiefly due to the self-sacrificing devotion of world-wide Christian Endeavor in contributing to the Headquarters Building Fund. There is not one country in the world that has failed to contribute to this fund. And from the day that the executive committee gave the order to the treasurer to secure $170,000, to finance the new building operations, down through the period of construction to the day of the dedicatory service, and the present moment, Christian Endeavor has stood loyally by its servants. To-day there stands on historic Beacon Hill, Boston, possibly one of the most unique buildings in the world, the World's Christian Endeavor Building, representing an investment of $220,000, fully occupied, safely rented, and bringing a net return of 4¼ per cent on the investment under present conditions. All of which is made possible by the liberality of an appreciating world-wide constituency.

Publishing Department.

As publication manager I would report a most unusual year in the department; unusual in that the epidemic of last fall, together with war conditions, created a degree of uncertainty among hesitating purchasers of religious literature. However, our business for the year ending May 31, 1919, equalled that of a year ago, but with a considerably reduced margin of profit due to the unparalleled increased cost of production, and our hesitancy in a proportionate increase in prices. The net profits for the year were $1,132.51. As a consequence we were literally forced to increase the prices of all literature and supplies, the same having gone into effect August 1, 1919.

The list of new publications produced by the department during the year number twenty-two leaflets, four booklets, one book, one new chart, and nine different kinds of novelties and supplies. We have in the hands of the printers now the "Endeavorers Daily Companion," "Prayer-Meeting Topics and Daily Readings," and "Junior Prayer-Meetings Topics and Daily Portion" for 1920. Also, two new books, one on the Quiet Hour and one on Junior Illustrations. This, together with fourteen different kinds of advertising matter, of which tens of thousands of copies have been distributed.

One regrettable feature connected with the work of the department is the loss of Mr. Walter R. Mee, who for twelve years was manager of our western office at Chicago. Mr. Mee resigned April 22, in order to accept a position as executive secretary with the Chicago Interchurch Federation. However, there is no loss so great but that it can be met, and this was clearly demonstrated when we secured the services of Mr. R. A. Walker, of Indianapolis, who for a number of years was the very efficient field-secretary of Christian Endeavor for Indiana. Mr. Walker has already proved his worth by his enthusiasm, tact, and business ability. His induction in office in the presence of Mr. Mee, and the officers of the Chicago and Illinois unions in attendance, was fittingly celebrated, April 22, by a dinner at Hotel Morrison, Chicago.

The publishing department during all the years of its existence has, and still is, meeting the purpose for which it was organized. First, for the publication of Christian Endeavor literature and supplies, to be used in the general diffusion of Christian Endeavor knowledge among a world-wide constituency, and, second, for the purpose of financing the parent organization in its work among young people throughout the world. During the past eight years the net profits of the publishing department were $61,539, of which every penny was contributed to the executive and missionary department.

The net assets of the department as of June 1, 1919, are $33,678.84. The future looks bright, hopeful, and encouraging.

My field-work during the year was in eleven States, covering conventions, conferences, summer schools, two colleges, and the Interchurch World's Conference at Cleveland. Miles traveled, 15,735. Attendance large.

When we review the work of Christian Endeavor, we cannot help noting its splendid achievements. Led by a magnificent corps of devoted and efficient State, field, and national secretaries, together with their associates, we say unhesitatingly and without egotism that there is no other religious organization in America to-day that is accomplishing, at so small an expense, so much constructive, instructive, and inspirational work in the development of young people as is State-wide and nation-wide Christian Endeavor.

Appreciating the privilege of working in Christian Endeavor, and with a spirit of thankfulness for your hearty co-operation, I respectfully submit this report.

REPORT OF KARL LEHMANN,

Southern-States Secretary, Made to the Meeting of the Trustees of the United Society of Christian Endeavor in Session at Buffalo, N. Y., August 5-10, 1919.

This report covers the two years from July, 1917, to July, 1919.

The past year especially has been a hard and trying one for us because of the "flu" and the fact that some 21,000 of our boys went to the colors, including our chairman, Lieutenant Curry, Field-Secretary Wyatt A. Taylor, and eight of our State Presidents. All this has made the carrying on and financing of the work very difficult for us.

In spite of all hindrances we have organized 912 new societies and added some 75,000 new members to our ranks in the South.

In addition to directing the work of the committee in our extension campaigns, planning itineraries for field-workers, getting out *The Dixie Endeavorer,* managing the supply business at headquarters, and any number of other things, the Southern-States secretary has travelled 42,560 miles, delivered 1,569 addresses to nearly 275,000 people, written more than 8,500 letters, and more than a million words for the religious and secular press.

On July 1 our committee had all bills paid and nearly $500 cash in the bank. Every Southern State except two reported all bills paid and money in the treasury on July 1, and those two said that plans on which they were working would pay them out by Sept. 1.

On July 1 we had in our office in Chattanooga a stock worth $864.46 against which we owed $505.77, on which we paid $335.82 during July, leaving $169.95 due on this stock which we are taking care of in regular payments as we sell goods. Mr. Shartle writes us that only the Chicago office is now selling more United Society supplies than we are selling.

During the year our paper has been issued regularly. We now print 4,000 copies a month, and have a paid circulation of about 3,600, 3,330 of which was received in new subscriptions during the year. The paper was issued at a net deficit of $67.97 this year, which was paid out of our all-South treasury. We have $100 a year in our budget for that purpose.

During the year we have supported three full-time secretaries, and had Miss Hooper, the national Junior superintendent, in the South for five months besides. Miss Hooper made such a fine impression in the South that we have invited her back for next year, and she has agreed to come to us from January 25 to April 1. Mr. Breg has also been constantly in the field in the South, working in Texas and supported entirely by the Texas Union. He is this year to have an additional worker with him in the field in Texas, and will have charge of the work of the new Southwestern Christian Endeavor Federation. We are adding an additional full-time field secretary in the person of Lawrence C. Little, of Winnfield, La., and besides Mr. Little, expect to have Mr. Brubaker, Mr. Walker of Illinois, and Mr. Spafford, of Michigan in the field for campaigns in the South. In all it is proposed to have ten field-workers in the South this year, including the native secretary, whom we support in China, and in addition to Dr. Poling and Mr. Gates, who have been invited for convention tours.

During the past seven weeks we have prepared and sent a special article on Christian Endeavor in Dixie to 135 newspapers and religious journals; many of them have published these articles giving Christian Endeavor much worthwhile publicity in the South.

For the past two years in the South we have set a goal of $15,000 each year for denominational missions. Both years we passed the goal, this year nearly doubling it. For next year the goal will be $25,000.

Plans for the New Orleans convention, the third All-South, are well under way, and July 8-11, 1920, are the dates set. It promises to be the greatest religious gathering for young people ever held in the South.

The proposed budget for the All-South Committee for the coming year is $7,650 and it is all in sight, including the $350 from the United Society. We wish to thank the United Society for the $300 given each of the first three years of our work and the $350 given last year. It is a much-appreciated contribution for this great extension campaign.

It is the seemingly unanimous opinion of the leaders in the South that the All-South Extension Committee should be continued beyond the end of the five-year period for which it came into existence and which ends in September, 1920.

May I add a personal word of appreciation to all the members of Board of Trustees and the officers of the United Society, especially Dr. Clark, Dr. Poling, Mr. Shartle, Dr. Shaw, Dr. Wells, Mr. Anderson, and others who have given us their generous encouragement in this work, and to the leaders in the South who have co-operated so heartily with us for the cause.

And now with this report I close four years of the work in the South. ten years of work with the United Society, which, by the way, began in this very city, Buffalo, N. Y., on September 14, 1909, when I met Dr. Claussen here and began a tour of this State, and fourteen years of strenuous work in the field for Christian Endeavor. They have been busy and often very hard years, to which I have given my very best. I only regret that I could not have given more.

I feel that the time has come for me to lay down this work, and consequently at this Conference I am asking the All-South Committee to release me. I shall be happy to remain long enough to help a successor gather up the details of the work, and then I shall be glad to step out after these fourteen years of service.

Respectfully submitted,

Karl Lehmann,
Southern States Secretary.

ANNUAL REPORT OF THE EDITORIAL SECRETARY OF THE UNITED SOCIETY OF CHRISTIAN ENDEAVOR, AUGUST 5, 1919.

To the Board of Trustees of the United Society of Christian Endeavor:

Dear Brethren:

The report of the editorial secretary which I offer to you to-day must necessarily be, in the nature of the case, a more or less drab recital of routine work done. It has been work for the Kingdom, however, and it has been done with joy.

In the past year I have written and put through the press the following booklets. First of all, in the fall of 1918, I wrote a twenty-four page booklet on "Christian Endeavor and the Sunday School," discussing the relation of Christian Endeavor to the organized class and the proposed departmental Christian Endeavor society. I wrote a "Christian Endeavor Year Book," giving a brief account of the work done in the States and containing much handy information about the various Christian Endeavor campaigns and standards. This book also serves the purpose of the Christian Endeavor directory.

I wrote "The Endeavorer's Daily Companion" for 1920. I prepared the topics for 1920, and these were submitted to the Young People's Interdenominational Committee and adopted by that body. I prepared the Christian Endeavor "Daily Readings," and also a four-page leaflet on "High

Points in Christian Endeavor History." In the fall I put through the press another book written earlier, "Christian Endeavor History Told in Brief." I prepared programmes for Christian Endeavor Day and Decision Day and for a union rally, as well as the outlines of activity for Christian Endeavor Week in 1919. I have also prepared and submitted an outline for the week's activities in 1920.

I wrote a full-page article on Christian Endeavor for *The Expositor* of last January, a 600-word article on Christian Endeavor and the work of reconstruction for the *Christian Herald*, and an 800-word article on the History of Christian Endeavor for a Christian Endeavor Bible. Besides this, I write for *The Boston Transcript* every two weeks a batch of Christian Endeavor items.

I have read and passed on several manuscripts offered to the United Society for publication, and I have put through the press, having prepared the manuscript and read the proofs of two booklets: "Your Questions Answered," by E. P. Gates and Karl Lehmann; and "Fifty-Two Varieties," by Harry Githens. We have now in the press a large book by Ella N. Wood, entitled "On the Highway," a series of blackboard and object talks that carry the Juniors through the whole Bible.

In the past year we have issued a number of leaflets the manuscript of which I have prepared and the proofs of which I have read. These include "The New-World Campaign," by Professor Wells; "Union Quiet Hour Superintendents," by N. A. Walker; "How to Start a Christian Endeavor Society," by Dr. Clark; a revision of the "Alumni Leaflet," by Dr. Poling; "Intermediate Christian Endeavor Societies," by Winifred Skinner and Paul Brown; "Christian Endeavor Principles," "Essential Committees," and "Officers' Duties Outlined," by Dr. Shaw; "Christian Endeavor and the Church," by E. P. Gates; "Where the Money Goes and How It Is Raised," by Dr. Poling. I read proof also of Miss Hooper's new Junior training chart, and of the Junior Topics and Junior Portion or Daily Readings.

The publication manager is issuing reprints of several chapters of Professor Wells's book, "Union Work." I have written a book entitled "Finding God in the Quiet Hour," and this will be published when funds are available. Much may yet be done for Christian Endeavor in this department when we are able to finance new ventures.

Respectfully submitted,

ROBERT P. ANDERSON,
Editorial Secretary.

CHAPTER XIX.

THE RESOLUTIONS.

Adopted by the International Conference of the United Society of Christian Endeavor, Buffalo, N. Y., August 10, 1919.

BY the grace of God Christian Endeavor enjoys continued and growing fruitfulness. In spite of the world war and its incalculable cost to the membership of our movement, and notwithstanding the great group of gold stars on Endeavor's vast service flag, none of our yesterdays have been as crowded with usefulness as every to-morrow is filled with promise. Humbly and gratefully acknowledging the fact that the Society of Christian Endeavor, following the life of our Lord, continues to grow in wisdom and stature, and in favor with God and men, we hereby pledge ourselves and our organization to renewed devotion to the will of the Master and to whatsoever He would like to have us do. "Trusting in the Lord Jesus Christ for strength," we promise Him that whatever may be our task and our society's mission, we shall always, when we see Him ahead or hear His call, say loyally, reverently, resolutely to each other and to all other Christian people about us, *"Come on, let's go."*

The restoration to health of President Francis E. Clark since the last International Convention, and his appearance on the platform and in the leadership of this Convention, are occasions for thanksgiving for world-wide prayer thus joyously answered.

But ours is not unmixed joy to-day, since General Secretary Shaw feels that he must surrender on January 1, 1920, the office which he has so long honored. The oldest United Society officer in term of service and one of the very richest in duty done wisely and well, his retirement will be deplored wherever Christian Endeavor is known and in all the borders of true Christian statesmanship.

But with the continuing counsels of both of these long-time executives new and numerous other practised workers are coming into the field. With promise of what hitherto has never existed,—something like adequate income for extension and development,—Christian Endeavor seems ready to follow Christ unto the uttermost parts of the earth, in the United States and Canada increasingly, among all the nations with which ours were allied in righteous demand for freedom, and as rapidly as may be, in the providence of God and with the co-operation of the Endeavor societies there, in those lands against which ours have warred, and all to the end that the Prince of Peace may reign in all hearts and beneath every flag.

President Clark's Message.

Pledging our unions and local societies to the programme so ably presented in the Convention message of our world-honored President, Francis E. Clark, we here and now adopt as the aims of Christian Endeavor for the next biennium the following programme:

Loyalty Campaign.

LOYALTY TO CHRIST.
LOYALTY TO THE CHURCH.
LOYALTY TO THE CHRISTIAN ENDEAVOR PLEDGE.
LOYALTY TO THE CHRISTIAN ENDEAVOR IDEALS.
LOYALTY IN SERVICE.

a. Pledged Service.
b. Outspoken Service.
c. Systematic Service.
d. United Service.

Our STANDARD of LOYALTY, "WHATEVER HE would like to have me do."

WHATEVER,
In Living, Doing, Being, Thinking, Speaking, Giving.
WHATEVER,
In Church, Sunday School, and Christian Endeavor.
WHATEVER,
In Home, Business, Community, and State.
WHATEVER,
In Recreation, Social Standards, and Amusements.

DEFINITE LOYALTY GOALS ON TWO-YEAR BASIS:

1. A Christian Endeavor society idea of fifty per cent net gain in membership.
2. One or more Christian Endeavor societies in every available church, or if not, the reason why.
3. A definite goal in every State and union organization for denominational missionary gifts.
4. A Sustaining Alumni Association in every union.
5. Development of union organization and Christian Endeavor fieldwork.

CHALLENGE, "COME ON"; Response, "LET'S GO."

The 1919 Committee.

This Convention, one of the greatest in practical results in our history, though by design limited in numbers and not assuming the proportions of a complete International Convention, such as it is proposed under stated conditions to hold in New York, 1921, owes deep appreciation to the Buffalo, 1919, Committee. Without the service of the general chairman and the tireless members we could not have had the comforts and conveniences that have made our own labors a delight. All have done for us whatever we could have expected, and some have anticipated our wants and sought to find ways to help far beyond any right we had to expect assistance. Because individual mention might seem invidious when nobody has left anything undone we do not name the persons who have blessed us since our varied excursion-leaders put us off at Buffalo; besides, time would fail us to mention those who denied us the privilege of mere human beings to find fault, for there were arrangements made for churches, halls, and decorations; for publicity and printing; for finance, and ushers, and entertainment and registration, and excursions, and reception, and billeting, and music and music's accompanying pianists, not to attempt to catalogue the varied attentions which no committee was named to pay, but which completed the cycle of our content.

The New Headquarters Building.

Since our last Convention the Headquarters Building of the United Society has been completed and dedicated in Boston. It is a most worthy monument to the long life of continuing service of the founder of Christian Endeavor, and a tribute to the like faithfulness of his co-workers now permanently housed therein and furnished for increased efficiency.

Affording commodious working-space not alone for United Society officers and their helpers, but providing a home for *The Christian Endeavor World*, without which no Endeavorer is informed and no society equipped, this building in the heart of Boston's noblest history and best modern life is a credit to every member who contributed to its erection.

Christian Endeavor and World-Betterment.

Recognizing the stewardship of public opinion of so vast an army of young people, Christian Endeavor has never shirked its duty to declare the whole counsel of God against public and private wrong and in behalf of the righteousness that exalteth nations no less than individuals. It denounced the legalization of the liquor traffic when to do so was an unpopular peril, and it was Christian Endeavor's that was the first voice that cried aloud for "A saloonless nation by 1920, the three hundredth anniversary of the landing of the Pigrims at Plymouth." Christian Endeavor protested that world's fairs should not open on Sundays, and Sunday openings became unprofitable. Christian Endeavor has steadily stood for good citizenship of all kinds, and the rights of every worthy American of both sexes to defend as electors the liberties and virtues of their children. Christian Endeavor steadfastly believes that a man should be first a Christian and then a political partisan when consistently he can be both, Being a world movement, it believes in international brotherhood, and seeks in international peace and good will. Therefore it deprecates and deplores any tendency in present-day politics that would make a partisan football of the bleeding heart of a war-wounded world, and it abhors the cupidity that would enrich itself unduly at the expense of the citizens of our own and other lands that offered everything, some of them giving everything, that the property and lives of this and unborn generations might be secure.

Specifically we are unalterably against, and to the extent of our force of millions of young voters and near-voters *pledge ourselves to prevent and destroy,*

Every popular prejudice, sectional, sectarian, partisan, and racial.

All forms of lawlessness and mob violence, whether its victims be individuals or races or classes, every such destruction of human life being the heinous crime of murder.

The alarming spread of such evil habits as profanity and cigarette-smoking, against which and numerous other practices that weaken influence and character we pledge ourselves to a campaign of education and the earliest practicable annihilation.

Anything in legislation or propaganda or practice that encourages an already dangerous disposition to regard Sabbath-desecration as a necessary evil, and we tender our aid to the Lord's Day Alliance in its beneficent crusade to make any necessary interference with the sacred observance of the Sabbath regrettable and any unnecessary desecration illegal.

All teachings or activities or governmental usurpations that countenance industrial injustice or commercial dishonesty, or that serve to separate the people into unfriendly classes and to widen rather than heal whatever breach may have existed between those who labor and those who pay for labor.

The clamor for war with Mexico, though we condemn the crimes against life and property which are fanning the wild fires of intervention. Millions for Mexican evangelization and education and ultimate good-neighborliness, but till every honorable other effort has failed not a dollar for slaughter. "The strong ought to bear"—not exploit—"the infirmities of the weak."

As strongly as we oppose these and other evils, *we stand for,*

A comprehensive social programme for the church to meet the social needs and to overcome the grave social perils of our young people.

The enforcement of law, whatever the law, that the evils of bad laws and the virtues of good laws may become immediately apparent.

The legal destruction of the beverage liquor traffic by every nation on earth, that there may be "a sober world by 1930."

The prompt suppression of the crime of war profiteering, and the adequate deterrent punishment of all those deliberately contributing to the high cost of living in a world impoverished and anhungered by war.

The utmost and most thorough evangelization and Americanization of new Americans, that henceforth neither internal paganism nor treason may menace democracy.

Continued interdenominational fellowship, and the ultimate union of closely related churches. Meanwhile we indorse the programme and high ideals of Christian co-operation proposed by the Interchurch World Movement, pledging to this movement the loyal support of Christian Endeavorers everywhere. We rejoice in the growing spirit of unity of which this movement is an exponent, and of which the co-operation of the Epworth Leagues of Buffalo have been to this Convention so delightful an expression.

The Thrift and Savings Department of the Federal Government at Washington and the organization of Saving Societies and the habitual purchase of Thrift and Savings Stamps, not alone that the Government may be aided, but that poverty and dependence may be avoided.

The League of Nations to Enforce Peace, with only such necessary reservations and interpretations as will not delay world peace nor postpone that great hour when selfish nations dare not unsheathe the sword to smite a neighbor.

The Boy-Scout Movement, with whose programme for manly strength and integrity we are in such hearty accord that we rejoice in entering upon a plan of the most intimate coworking, coveting the closest possible co-ordination.

Finally, though we recognize the foregoing as by no means a comprehensive statement of our views nor of the questions of the day with which we are at variance or in harmony, we especially favor such a campaign of education and regeneration of both whites and blacks as shall result in the mutual helpfulness of both races, bound as they are to an existence of neighborhood. We denounce without measure such shameful riots as have lately dishonored both races and stained the flag of the republic. For these horrors we believe that the worst and most ignorant, and not the best and wisest, of both races are to blame; and that without fear or favor all the criminals should be brought to justice. But the terrible disease is too deep and widespread throughout the nation for local treatment or for social and political panaceas; only the truths of Christianity and the spirit of Christ can cure this cancer. Christian Endeavor offers fearlessly, immediately, and with faith in God, to apply its considerable share of the only healing remedy.

Respectfully submitted,

IRA LANDRITH, *Chairman;*
FAYE STEINMETZ, Oregon;
E. T. FARRILL, Wisconsin,
Committee.

CHAPTER XX.

SECRETARY SHAW'S RESIGNATION.

24 July, 1919.

President Francis E. Clark,
and Members of the Board of Trustees of the United Society of Christian Endeavor.

My Dear Friends:

Thirty-six years ago Christian Endeavor found me in my young manhood, and as a member of the Christian Endeavor society in the Phillips Congregational Church, South Boston, of which Dr. Clark was pastor, I received my training in religious work. Having left school when I was twelve years old, Christian Endeavor became my school, my college, and my university.

Three years later I was elected treasurer of the United Society of Christian Endeavor, and in the same year was induced to leave my business and become advertising manager of *The Golden Rule,* now known as *The Christian Endeavor World.*

In 1889, after three years of service with the paper, I was elected manager of the United Society Publication Department, which had been developed by The Golden Rule Company. I continued in this position until 1898. During these nine years of service the business increased until it provided all the funds expended by the United Society in the development and extension of Christian Endeavor. When I resigned, the bills were all paid, and we had a reserve fund of more than $18,000 in the treasury.

In 1898 I was called back to The Golden Rule Company, this time as business manager of *The Christian Endeavor World,* which position I held until 1906, when I was elected general secretary of the United Society of Christian Endeavor, succeeding Mr. Vogt. At this time I resigned my position as treasurer, which I had held for twenty years.

In 1910 I was again appointed business manager of *The Christian Endeavor World* in connection with my work as general secretary. Each time that I took charge of the paper it was in financial difficulties, and each time I had the satisfaction of seeing it restored to a sound financial condition.

It has been my joy to see Christian Endeavor grow from less than fifty societies, with 2,500 members, to its present world-wide constituency of more than 77,000 societies, with nearly 4,000,000 members.

No man ever had more considerate or helpful associates than mine have been, and no man ever enjoyed his work more than I have.

Next year I shall be sixty years young, but in spite of the protestations of my closest friends, too old, I am convinced, to remain longer as general secretary of the United Society of Christian Endeavor. The time has come to commit this task to younger hands. I therefore present my resignation as general secretary, to take effect January 1, 1920, or earlier if my successor can be secured.

I came to Christian Endeavor with the enthusiasms and visions of youth. I have given to it the loyal service of my life. When I came to it, it was small in numbers and without any financial resources. I leave it large in numbers and valued at a quarter of a million dollars, and loyal alumni, who, I am sure, will see that it does not lack for adequate resources to

Dr. Francis E. Clark,
Speaking at the Canadian Rally, Niagara Falls.

meet the challenging opportunities of the new day that is just dawning on the world.

Words cannot express my sense of indebtedness and gratitude to my associates for all that they have been to me, to the young people for the inspiration they have supplied, and to my heavenly Father for His sustaining grace during all these years of blessed service. I can only say, "Thank you," and assure you that my interest in and service for Christian Endeavor in other ways will continue until death opens wide the door into larger life.

Fraternally yours,
WILLIAM SHAW.

The Trustees' Resolution Concerning Secretary Shaw's Resignation.

WHEREAS, Our beloved general secretary, William Shaw, finds it necessary because of health considerations to resign his office as secretary of the United Society of Christian Endeavor,

Therefore, *Resolved,* That we hereby express to the world our testimony to the invaluable contribution which Dr. Shaw has made to the great movement of Christian Endeavor in particular and to the cause of Christianity in general. None has exceeded the tireless devotion and unselfish consecration with which he has worked in the interest of the young people of the world. The magnitude of the Christian Endeavor movement to-day is due in a large degree to his splendid leadership. Through his efforts thousands of young people have been trained for responsible places of service in the kingdom of Christ. For a generation he has been developing Endeavorers who will now gladly take up the responsibilities imposed by his resignation. His work has been well done, and will go on, a perpetual testimony to his spirit and ability.

To Dr. Shaw we can give no adequate expression of the strength and tenderness of our affection for him. Millions of Endeavorers throughout the world will feel a deep personal loss because of his official retirement. We have felt the touch of his great personality, and are better because of it. As a heritage of our organization we will hold in perpetual appreciation his unparalleled service to our common cause, and here pledge ourselves in renewed consecration to the successful consummation of the tasks to which he has inspired us.

CHAPTER XXI.

THE CLOSING SESSION.

ELMWOOD MUSIC HALL.

Sunday Evening, August 10.

On the Mount of Vision.

LONG before the doors of Elmwood Music Hall opened for the closing session of the Conference on Sunday evening, crowds of Endeavorers were already in front of the hall, and a steady stream of young people poured through adjoining streets. Reservations in the hall were made for the various State delegations. Before the fine organ recital by Mr. George Bagnall was finished, the auditorium was filled, and the gallery as well, and people were standing at the back of the hall.

The choir on this closing night did especially fine work, and George E. Gowing, the leader, who proved himself a master throughout the Convention, was at his very best. Indeed, the whole audience was a choir, singing in glorious harmony. It was thrilling, absolutely soul-stirring, to hear these thousands of fresh young voices sing the praises of the King. When the leader asked the men in the house to stand, it was a small army that rose to sing "Brighten the corner." The young women had their chance also to "Brighten the corner," and they did. Then all the visiting delegates were given their opportunity, and they sang with magnificent force and spirit. The whole spirit of the Convention was put by the audience into "Onward, Christian soldiers." The choral may have been sung as well before, but never better. It was given with one mighty voice, the voice of youth and of faith, and the effect was thrillingly overpowering.

The devotional service was conducted by Rev. A. W. Bourne, D.D., a Buffalo pastor.

The Convention sermon, "Back to God," was to have been preached by Hon. William Jennings Bryan, but, to the disappointment of all, Mr. Bryan could not be present, because of the illness of his wife. He had sent his address, however, and although the great audience could not look into his face and listen to his voice, they had the privilege of hearing his message. Mr. Bryan's speech was read by Dr. Poling, and appears in full in this report.

Dr. Poling announced that the first collection of the Convention would now be made in order that funds might be provided to purchase a suitable gift for General Secretary Shaw on his retirement. No man has served Christian Endeavor more wholeheartedly and more generously than Dr. Shaw, and it seemed fitting that the Endeavorers of the country should in some tangible way express their appreciation of his magnificent service. The joke of the evening was that Dr. Shaw, who would not have permitted a collection for this purpose, was cunningly sent to look for Dr. Ira Landrith, who was to read the resolutions of the Convention; and while Dr. Shaw was outside on his vain search—for Dr. Landrith was not lost by any means—the offering was made. Dr. Clark was asked to be chairman of a special committee to arrange for a suitable gift. One may imagine the delight of the audience and the surprise of Dr. Shaw when the little trick was revealed to him.

Dr. Shaw, of course, had to respond, and made one of his characteristically bright speeches, a speech, however, that soon swept into serious channels, and became a moving appeal to the Endeavorers to do something for the foreign work of Christian Endeavor rather than for himself. Twenty dollars keeps the world-wide work of Christian Endeavor going for a whole day, or trains in Christian Endeavor 275 young people in foreign lands for a whole year. Dr. Shaw asked those who were willing to make such a gift to stand, and a host of young people rose to their feet.

Dr. Ira Landrith broke into the programme at this point and read one of the resolutions prepared by the resolutions committee, a resolution brimful of affection for Dr. Shaw, and the Convention indorsed its sentiments by a standing vote.

With ringing voice Dr. Poling then read the principal parts of Mr. Bryan's splendid address, to which the audience listened with rapt attention. The audience voted to send to Mr. Bryan a telegram expressing the sympathy of the young people, and Dr. Clark offered a brief prayer for the recovery of Mrs. Bryan.

Dr. Clark introduced Dr. Ira Landrith as one of the very few men in the country who can take the place even of an orator like Mr. Bryan, and the audience not be disappointed. Dr. Landrith, disclaiming the ability to take Mr. Bryan's place anywhere, paid a warm tribute to the great Commoner, and then read the resolutions prepared by his committee. These resolutions will later be published in *The Christian Endeavor World*. It is interesting to note that the last time Dr. Landrith read resolutions at an International Christian Endeavor Convention was at Atlantic City in 1911, when he proposed the great Christian Endeavor slogan, since gloriously realized, "A Saloonless Nation by 1920." This time he proclaimed a saloonless world by 1930. Dr. Landrith's pungent language and red-hot indignation at all forms of iniquity disclosed whose hand wrote these resolutions. The resolutions were not mere glittering generalities, but came down to vital particulars. The opinion of the young people was expressed throughout the reading by bursts of delighted applause, and at the close by a rising vote. Dr. Landrith's "asides" and comments were gems. Probably there is no wittier or pithier speaker on the lecture platform in America than Dr. Landrith, who amply justified Dr. Clark's introductory words of praise.

Next came one of the most interesting features of this Conference, "an interdenominational love-feast," a gattling-gun, rapid-fire series of three-minute talks by representatives of more than twenty Christian Endeavor denominations. It was a great occasion that demonstrated practical Christian coöperative union, and proved that the hearts of the churches stand for love and service and mutual esteem.

The love feast began by the presentation to the audience, by Dr. Poling, of those officers of the United Society who have been elected since the last Convention. Dr. Poling called them to the platform, one by one, and they formed a splendid group, drawn from the Christian Endeavor forces of many States. They comprised E. P. Gates, field-secretary; Paul Brown, Intermediate superintendent; Grace F. Hooper, Junior superintendent; Stanley B. Vandersall, Alumni superintendent; Rev. Robert P. Anderson, editorial secretary; R. A. Walker, Chicago representative; W. Roy Breg, Southwestern secretary; and Clarence C. Hamilton, field representative of *The Christian Endeavor World*. Treasurer Shartle was introduced later in the evening and spoke a few words.

The love feast proper began when Rev. Ralph W. Brokaw, D.D., for many years a beloved trustee of the United Society, spoke for the Presbyterian Church, U. S. A. He agreed with Dr. Landrith's great speech, and declared that friendship should be the key-note of our movement. God is the Friend of all the world, and it is the business of the church to put men into friendly relations with one another. The whole world yearns for friendship. The Presbyterian church has a warm heart as well as a clear intellect, and it welcomes to pew and pulpit all who love our Lord.

Paul C. Brown, California's field-secretary, spoke for the Congregationalists. The idea of fellowship in Christian service, he said, expressed now in the Interchurch World Movement, was born of the Congregational heart. Never for a moment in all its history has Congregationalism proved untrue to this splendid ideal of fellowship so beautifully expressed in Christian Endeavor.

H. N. Lathrop, clerk of the United Society, speaking for the Baptists, told of the $100,000,000 campaign of the denomination, working on a five-year basis, and referred with warmth and feeling to Dr. and Mrs. Clark, who have been father and mother to him, and to his fellow officers in the United Society office, where the workers, representing a number of denominations, have a perpetual love feast. The Baptists believe in interdenominational fellowship, and call now to all, "Come on!"

The African Baptist churches were ably represented by Rev. W. T. Johnson, D.D., a trustee of the United Society. He spoke warmly of the part the colored church is to play in the $100,000,000 campaign. They are to give $25,000,000 of this amount. Dr. Johnson added, with reference to Dr. Landrith's fiery plea for fair play and a square deal for the Negro, that the Negro delegates would go back home greatly encouraged and strengthened by the sympathy of the Endeavorers.

Rev. George Strayer represented the United Brethren Church. He was glad, he said, to speak for the United Brethren in Christ, a small body of 350,000 members; but after attending the Convention he was ready to believe that we are all united brethren in Christ. Speaking for his denomination, he pledged the young people to the great programme of the United Society for loyalty and coöperation, and added:

We'll tell it wherever we go,
We'll tell it wherever we go,
That Christian Endeavor will live on forever,
We'll tell it wherever we go!

The Reformed Church in the United States was represented by Rev. Rufus W. Miller, D.D., a trustee of the United Society. He said that his denomination believes in the programme of the United Society as it has been developed at the Conference. This Conference, he said, is one of the most practical that has ever been held in Christian Endeavor history. The Reformed Church believes in progress, therefore it hails the Interchurch World Movement. It believes in organized union. It stretches out the hand of fellowship to every denomination. It looks ahead, and its call is the Christian Endeavor slogan, "Come on!" Dr. Miller closed with putting the slogan into an acrostic:

Christ's
Order:
Move
Enthusiastically,
Obediently,
Now.

Dr. Gilbert Glass said that his denomination, the Southern Presbyterian, is "restricted geographically but not in spirit; its arms of affection are long enough to reach every one of you." He spoke of the fine leadership of Christian Endeavor in the nation and States, and of the tremendous task that lies before the Society. "Our church is with you, hand to hand, shoulder to shoulder, heart to heart."

The new young people's secretary of the Christians, Rev. A. B. Kendall, told us that, though his denomination "would not add much numerically to the Christian Endeavor society, the spirit and principles of Christian Endeavor are identical with those of the Christian Church. By the grace of God we are going to do our part to put over the magnificent programme adopted by this great Convention."

That splendid field-secretary, Charles Evans, spoke for the Disciples of Christ. "Christian Endeavor is peculiarly appealing to my denomination," he said, "because of its interdenominataional spirit and its emphasis upon the Bible." He told of the largest Christian Endeavor society in the world, that of the Disciples mission in Bolengi, Africa, with 1,900 active members. In the United States and also in the South the Disciples are second in the number of societies. "We are coming more and more," he said, "to apprecite this great opportunity of training for Christ and the church."

Dr. Joseph B. Baker speaking for the Lutherans, pointed to this series of concise less-than-three-minutes speeches as evidence that pastors know how to stop. He told the story of two hoboes who paused in their track-walking to request a telegraph operator, "Comrade, suppose you report two empties going east." "But there will be no empties," Dr. Baker continued, "going north, south, east, or west from this great Convention." He told the striking story of the murderer who witnessed his own funeral, walked forward in protest, and saw *himself* lying in the coffin. "Endeavorers," said Dr. Baker earnestly, "do not go home and murder yourselves, those glorious new selves that were born here in Buffalo. Billy Sunday says that many Christians sing, 'I'll go where you want me to go, dear Lord,' and then go to bed. Endeavorers, go home and *go to work*."

The African Methodist Episcopal Zion Church was represented by Mr. Lloyd Mitchell, of New York City. He spoke of the 400,000 members of his denomination, most of them enthusiastic Christian Endeavorers. "During the next two years," he promised, "we are going to have a Christian Endeavor society in every one of our churches, or know the reason why."

General Secretary Peffley, head of the young people's work in the United Evangelical Church, expressed the gratitude of his denomination for what Christian Endeavor has enabled it to discover and develop. In Poling and Vandersall it is paying back a part of the debt. He spoke of churches in the West and in China which United Evangelical Endeavorers have built, and said that in the China mission no one is admitted to church-membership that has not received the Christian Endeavor training.

Rev. Ora W. Carrell, trustee for the Friends, spoke of his people as one of the smaller bodies with a very old history. Young people are always given a large place in its service, and

Christian Endeavor is their largest organization for young people. "It is their earnest desire to be Friends of the truth as it is found in Jesus Christ, and Friends of all mankind."

Speaking for the Seventh-Day Baptists, Rev. William M. Simpson said he supposed there were not ten thousand of them in all the world, "but because we are so few we do not propose to withdraw from the fellowship of Christians the small amount of power we have. We are glad to join with other denominations through Christian Endeavor."

Dr. John F. Cowan, writer of our Sunday school page, was warmly greeted as he came forward to speak for the Methodist Protestants. "Christian Endeavor is the only brand of young people's societies in our branch of Methodism," he said, "and we find that Christian Endeavor makes our young people as loyal as any other type of society possibly could."

Minnesota's fine field-secretary, Miss Margaret M. Bradt, represented the Episcopalians. She hoped to see "Episcopal succession turned into the Christian Endeavor procession," for when her great church goes into anything it goes into it heart and soul. She promised to work for a Christian Endeavor society in every Episcopal church in Minnesota.

The speaker for the Cumberland Presbyterians was another of our splendid young women, Miss Emma McClaran, president of the Oklahoma union. "My denomination," she said, "which is especially strong in Tennessee, Texas, and Oklahoma, intends to stand 'for Christ and the church.' We are growing and we are back of Christian Endeavor."

The spokesman for the Church of God was its trustee, Rev. T. M. Funk. He was glad of broadening denominational vision, especially of denominational membership in the Federal Council of Churches. Their denominational programme includes a 30-per-cent increase each year for eight years.

Ohio's capital Junior superintendent, Miss Sylvia L. Tschantz, represented the Mennonites. She spoke of the history of her church, reaching back to the beginning of the Reformation. "We are not many, only 20,000 in all, and not all of our churches have Christian Endeavor; but those that do have are proud to belong to the great army."

Amos R. Wells, called upon unexpectedly to close the "love feast," told how Christian Endeavor had made him a member of the Presbyterian church after he had spent years in making false excuses for remaining outside the church; and how, going on from that, Christian Endeavor had made him a member of sixty denominations. For this is the spirit of Christian Endeavor, that each of us, earnestly loving and serving one denomination, is bound none the less but all the more to the whole Church of Jesus Christ.

Field-Secretary Gates then made the closing announcements, including the Conference registration. There were 2,083 delegates, representing forty-four States, two Provinces of Canada, and five foreign countries. In addition, seven foreign countries sent greetings.

A great surprise was an earnest word of greeting from Rabbi Cantor, associate of Rabbi Wise of New York City, who had sent his card to the platform and was asked to speak to the Endeavorers. He was received with overwhelming applause as he came forward and made an eloquent plea for justice to the Jew. "Israel Zangwill," he said, "has written that for two thousand years Christianity has preached that when men strike us on one cheek we must turn the other, and Judaism has had to practice it. Now in this new era of reconstruction let there be a reconstruction of the Christian spirit, so as to permit full and free sympathy with suffering people everywhere."

The closing words of the Conference were spoken impressively and inspiringly by Dr. Clark: "More than once the speakers of this evening and during the Conference have declared that they have had a vision. It has been a vision of brotherliness, of service, of fellowship, of the future that awaits us. Shall we lose this vision? Pray God that we never shall. Is it necessary ever to go down from the mount of vision into the valley? No, never. Our visions are not a pinnacle far up in the skies where we can stand for only a few minutes, but a plateau, a tableland, where we can stand all our lives. Do not think of going down, of receding from the high places that you have reached. Think of this experience as the normal state, the place where the Christian may live.

"President Coffin of the New York State union, said the other day to the newly-elected officers, 'You will either be a channel of blessing to others, or you will choke up so that they cannot receive the blessing through you.' Thus also every attendant at this Conference will either be a channel through which the blessings of these meetings will flow to others, or so choke up the channel by negligence and thoughtlessness that the blessing will go no farther. No one can report this Conference except those that have been here."

Dr. Clark then read Paul Brown's fine arrangement of the goals for the coming two years which were proposed by the president of the United Society, the audience repeating each one after him. Then they joined earnestly in the motto of the two years, the "whatever" clause of our pledge, and ended with the slogan: *"Come on!"* cried Dr. Clark. "LET'S GO!" shouted the Endeavorers with a will.

The beloved founder of Christian Endeavor offered a tender prayer as we all stood, pronounced the benediction, and the Buffalo International Christian Endeavor Conference passed into history.

CHAPTER XXII.

MESSAGES OF GREETING.

A Message from Europe.

[Rev. John Pollock, president of the European Christian Endeavor Union, had planned to be present at the Conference, and had secured his passport; but it was absolutely impossible to secure transportation of any kind. He sent the following message.]

15 July, 1919.

To the Delegates Asssembled in Conference at Buffalo, U. S. A.:
My Dear Friends:

Let me assure you at the outset that I very highly appreciate the courtesy of an invitation to your fellowship and to participate in your deliberations. I accepted the invitation with genuine pleasure, and very deeply regret my inability to be present on an occasion to which I had looked forward so eagerly. Especially should I have esteemed it a distinguished honor to convey to you in person the greetings of the European Christian Endeavor Union, and to join with you in thanksgiving to Almighty God for the blessing of peace restored.

Permit me to remind you that I do not address you as a British Endeavorer or as a British citizen, but as representing the Endeavorers of Europe on both sides of the firing line, now happily silent. The desire of the European Council—those members of it with whom I have been able to maintain correspondence—is that Endeavorers in all the European countries may speedily be brought to the fuller recognition of the common Fatherhood of God, the fundamental brotherhood of man, and, above all, the catholic unity of the body of Christ; and that Christian Endeavor may have a large and honorable part in the great and urgent task of closing the world's wounds. As one of the wise has well said, "The sword makes war; the heart alone can make peace."

Doubtless you have already formed some estimate of the difficulties confronting us in planning and prosecuting any practical policy of reconciliation and reconstruction. It is to be hoped that the way may soon open for the admission of the Central States of Europe into the League of Nations. This is obviously essential to the effective influence of the League in bringing about the realization of the best aspirations of mankind, the fulfilment of the brightest dreams of sages and seers. It is felt that the ratification of the treaty of peace is not sufficient guaranty of success in any effort our union might make to bring together again the Christian Endeavor forces of Europe. We prayerfully await the opportunity which the completion of the League will provide.

You will be glad to know that in central Europe Christian Endeavor has maintained a vigorous existence during the war. Even in Germany the movement is holding its own. We long for the restoration of those relations of cordial friendship and affection by which the Endeavorers of Europe were bound together, and which found expression in most memorable European conventions.

May our good cause, beloved on both continents, flourish in all the lands represented at your Conference. And may the Spirit of wisdom preside over all your meetings.

"The grace of our Lord Jesus Christ, and the love of God, and the communion of the Holy Ghost, be with you all."

Fraternally yours,

JOHN POLLOCK,

President of the European Christian Endeavor Union.

LINCOLN, NEB.

MY DEAR MR. SHAW:

Much to my regret, Mrs. Bryan's ill health prevents my coming to the Buffalo Conference. I am greatly disappointed, for I have a message for the Christian Endeavorers, whose service of Christ, the church, our country, and the world I highly appreciate; and I was anxious to deliver it at Buffalo.

Give my regards to Dr. Clark.

WILLIAM JENNINGS BRYAN.

SHANGHAI, CHINA.

Greetings from your forty thousand Chinese fellow Endeavorers. Outlook for Christian Endeavor in awakened China was never more hopeful. Pray that China's Endeavor societies may be increased several-fold. "Pray ye therefore the Lord of the harvest, that he will send forth laborers into his harvest."

EDGAR E. STROTHER.

From Mexico.

OUR BELOVED BRETHREN:

The Mexican Endeavorers send a kindly and joyful salutation to all their brethren and comrades gathered at Buffalo. May God bless you all and lead in all your deliberations and resolutions!

Three days ago we had an interesting National Convention of Christian Endeavor at the capital of one of our States. We saw and felt clearly the influence of the Holy Spirit of our heavenly Father in all our meetings.

May the Holy Spirit bless you!

ALFONSO PINERA GALLARDO,

General Secretary, Mexico City, Mexico.

Finland.

The Finnish Endeavorers send hearty greetings to their American brethren. After four terrible years of war from beyond their boundaries and civil war within their borders the forces of Christian Endeavor have gone forward in their work for the Master. They find there are 30 Finnish-speaking societies and 27 Swedish-speaking societies in Finland, with 2,006 active members.

Jugo-Slavia.

The Endeavorers of Jugo-Slavia send their greetings. The first national convention of the societies will be held in the fall. They send hearty good wishes to their friends in America.

South Africa.

The Christian Endeavor Union of South Africa sends warmest Christian greetings to the International Convention at Buffalo, New York.

"We as servants of Christ Jesus salute you, always striving for you in our prayers that ye may stand perfect and fully assured in all the will of God." Col. 4:12.

REV. A. E. BLISS, D.D.,

Honorary President, South African Christian Endeavor Union.

JULY 4, 1919. *Wellington, South Africa.*

New South Wales.

Endeavorers of New South Wales, Australia, send hearty greetings to the Buffalo Christian Endeavor Convention. The comradeship of former years has been deepened and strengthened by a fellowship in common suffering for the cause of freedom and righteousness. As we write, we are rejoicing in the celebration of peace. We are planning our State convention for October 9 to 14, with "The Supremacy of Christ" as the theme. "Unto him that loved us and washed us from our sins in his own blood, and hath made us kings and priests unto God and his Father; to him be glory and dominion for ever and ever. Amen." Yours in prayer for a mighty spiritual blessing upon your Convention.

REV. W. P. PHILLIPS,
President.

Chicago.

Chicago Conference.

Kindly accept the congratulations of the Chinese Christian Endeavor society of Chicago. Though we are unable to attend the Conference, our heart and prayers are with you. God will abide with you in working among the young people, upon whom the realization of the brotherhood of men depends.

MRS. P. S. CHAN,
President.

WILLIAM SHAW:

Unexpected duties here arising from our peculiar position as a city prevent my coming, greatly to my regret. Our whole church with me send heartiest congratulations on the splendid advance Christian Endeavor is making through its consecrated leadership, and offer fervent prayer for still more glorious achievement in coming days.

SAMUEL FALLOWS.

Maritime Provinces.

DR. WILLIAM SHAW:

Greetings to Dr. Clark and the officers of the United States societies from the young people of the Maritime Provinces of Canada. Forgetting the sad past, we advance with you, filled with home and expectation. Our motto, the motto of our soldier Prince of Wales, "I serve."

W. H. BARRACLOUGH.

Burlington-Graham, N. C.

KARL LEHMANN:

Convey to Convention fraternal greetings and assurances of co-operation from Burlington-Graham, N. C., union.

J. O. CORBETT,
President.

LONDON, ENGLAND.

REV. FRANCIS E. CLARK:

I earnestly pray that God's richest blessing will rest upon you and all who assemble at Buffalo, and that this great Convention will be one with far-reaching results. Every good wish and fraternal greetings from British Endeavorers.

JAMES KELLY,
Field Secretary of the British Christian Endeavor Union.

DES MOINES, IA.

WILLIAM SHAW:

On behalf of the people of this commonwealth I urge favorable consideration of our invitation to hold your 1921 Convention in Des Moines, Ia. Would feel honored in being host to this gathering.

W. L. HARDING,
Governor.

PAUL C. BROWN:

Your executive unanimously indorses Loyalty Campaign for California. Let's go! Our greetings to the Conference.

BENJAMIN SMALL.

The President's Response to the Buffalo Resolutions.

Dr. Clark sent to President Wilson at the request of the Buffalo Conference the resolution unanimously passed in favor of the League of Nations and against intervention in Mexico, except as a last resort, after all diplomatic methods had failed. Although the letter called for no acknowledgment the President took time on one of his busiest day, to send the following autographed letter, thus showing how much he appreciates the support of Christian young people on this momentous issue of the League of Nations.

The following is his letter:

THE WHITE HOUSE, WASHINGTON,
21 August, 1919.

MY DEAR DR. CLARK:

You know how to cheer me, and the information brought me by your kind letter of August 12 has indeed done so. I thank you with all my heart.

Cordially and sincerely yours,
WOODROW WILSON.

CHAPTER XXIII.

GLEANINGS.

The Christian Endeavorers had a great week in Buffalo, and the closing meeting of the Conference was a fine testimony to the worthiness of the movement, taking in, as it does, all Protestant denominations that are willing to join this great society of service. Mr. Bryan's absence was not permitted to dampen the enthusiasm of the gathering at which he had been expected to speak yesterday.—*Buffalo Express.*

This is a post-war Conference of the United Society of Christian Endeavor. The war and conditions growing out of it only have served to emphasize the plan of action laid down as long ago as 1893. This provides for practical service to the country and emphasizes the importance of Christian citizenship.—*Buffalo Evening News.*

"I hear a multitude of voices calling—I see a sea of hands raised against the recurrence of war. They want us to adopt a League of Nations which will be consolidated on the basis of the brotherhood of men and the Fatherhood of God," said Secretary of War Newton D. Baker in his address to the International Christian Endeavor Conference last night. He looks forward to the future of the United States with confidence, he declared. The hall was crowded. Many persons were standing.—*Buffalo Courier.*

Henry S. Gage, general chairman of the Buffalo Conference committee; George J. Barnett, chairman of the committee on publicity, and Walter F. Alt, chairman of the committee on excursions, said last night that they would like to have it known that Secretary Baker was the guest of the Christian Endeavorers; that they had secured his presence for the Convention, and were entertaining him at the Hotel Lenox. They did not begrudge the secretary's address before the Knights of Columbus, also in session in Buffalo, but wanted it clear that the secretary was the Endeavorer's guest.—*Buffalo Express.*

No one could have a more pleasant face than that of the Rev. Dr. Francis Edward Clark, founder and world president of the Christian Endeavor Society. His gray eyes have the look of the father of a large and happy family, and about them are the wrinkles that come from smiles and laughter. He wears a mustache, which is gray, and, though he is partly bald, he has gray hair. His face is browned from outdoor life and work on his Cape Cod farm, where he spends two or three days every week. But what one notices first and most in Dr. Clark's appearance is the smile that lurks ready around the eyes and corners of the mouth, where, seemingly, it has appeared so often that one can see it when it really is not there. Dr. Clark is most unassuming for the head of so great an organization.—*Buffalo Express.*

Delegates from many parts of the world are in attendance at the Christian Endeavor Conference. They come fresh from great accomplishments at home and abroad. They plan for still greater activities now that the world is ready for the ways of peace. They have vision and they have that spirit of helpfulness that will make their visions come true for the betterment of their fellows.—*Buffalo Express.*

The literature tables, which were in charge of R. A. Walker, formerly field-secretary for Indiana, now manager of the Chicago office of the

United Society of Christian Endeavor, were well filled with all sorts of Christian Endeavor helps. The Endeavorers took advantage of the opportunity of looking over the many helps that are at their disposal.

Delegates found the members of the Convention committees splendidly ready to render whatever service was desired.

The Des Moines, Ia., delegates, who came to invite the next International Convention to their city, looked like Christian Endeavor "Reds," or, as some called them, Christian Endeavor Bolsheviki. The ladies wore flaming red blouses, and the men equally flaming shirts. They were happily peaceful, however, in all their ways.

Delegates from St. Louis who came to invite the 1921 Convention—the fortieth anniversary of the founding of the first Christian Endeavor society—had with them a liberal supply of large buttons which showed a red map of the United States with the city of St. Louis in the centre. When making his plea to get the next Convention one of the delegates claimed that St. Louis is the only large city in the country that literally is bounded by the United States.

At the back of the great auditorium Oregon had stretched across the front of the gallery an immense banner with the words, "Portland, Oregon, 1923." Miss Steinmetz, president of the Oregon union, served notice in a meeting of the trustees that Portland wishes the Convention in 1923, and to point out that the city has asked for it well in advance of the time.

The Convention badge was a round button, a little larger than a quarter-dollar, bearing a buffalo's head with the Christian Endeavor monogram on its forehead. A ring of gold ran around the rim of the button, and inside this ring were printed in red letters the words, "International Conference, Buffalo, N. Y.," the dates August 5-10, being printed on the white ground inside the ring. To the button was attached a piece of ribbon of the Christian Endeavor and national colors, red, white, and blue. The effect was striking and beautiful.

The choir at the Buffalo Conference was a small one, as convention choirs go—only about 150 members—but the singing has never been surpassed in quality in any Convention. The members proved faithful, too, at all the great sessions. Altogether, the singing did great credit to both the choir and to its leader, George E. Gowing.

"We're from Ohio"—at least, one of us editors is; and we were especially glad of the distinction when honored by an invitation to the Ohio banquet. It was snappy from beginning to end, and enjoyed speeches by leading Ohio Endeavorers as well as the four national officers who are "from Ohio."

One of the delegates was an Intermediate girl from Pine Bluff, Ark. She was born in Switzerland, but came to this country ten years ago. She is now in high school, and wears a diamond medal given by the W. C. T. U. for elocution. She is a member of the church, and is active in its work.

During the Conference the All-South Extension Committee held a meeting at which were made reports of officers and departments in Dixie. Lieutenant Duncan B. Curry, of Jacksonville, Fla., was re-elected president; the secretary is Wyatt A. Taylor, Columbia, S. C., and the treasurer, Mrs. Wyatt A. Taylor. Karl Lehmann, Chattanooga, Tenn., was re-elected Southern-States secretary. The general field-secretary is Charles F. Evans, Lexington, Ky., and L. C. Litttle, Winnfield, La., is the field-secretary.

The Conference, with its more than two thousand registered delegates, was much larger than the officers had expected, and crowded the biggest auditorium in Buffalo. Christian Endeavor could not undertake one of its usual vast Conventions in this year of reconstruction and turmoil; but the Conference that was undertaken, though not in our Convention series, really belongs there in numbers as well as in power and influence.

It was a youthful crowd, alert and jovial, sparkling with all of youth's enthusiasm, responsive to all fine impulses, fired with eager resolves, a most lovable and hopeful assembly. The singing, both of the choir and the audience, was superb. The cheers and impromptu songs, of which so much has been made in recent State conventions, burst our irrepressibly whenever a good opportunity was afforded. No one could be in Buffalo and have any doubt of Christian Endeavor's abounding vitality and undying zeal.

It was a faithful crowd. The Endeavorers did not go to Buffalo to see the sights, though there were many sights well worth seeing, and they came from distant parts of the country. They went there to attend the Christian Endeavor Conference, and this they did with constant and glad fidelity. Not only were the evening mass-meetings all crowded, but so also were the more practical and less inspirational conferences of the daytime. Nor were the Endeavorers satisfied with the crowded Conference programme, but they packed the time between sessions with impromptu conferences on all sorts of helpful themes, till we wondered when they ate and slept.

The Buffalo Conference also brought out in splendid fashion the quality of Christian Endeavor's leadership. It showed for the first time the new personnel of the national officers, the able and winsome young men and young women who have been drafted during the past two years into the forces of the United Society from the State unions. It also made evident the vigorous and delightful personalities of the State officers, including the constantly enlarging number of field-secretaries. Christian Endeavor has never before been so well manned—and womanned.

The attitude of the denominations toward Christian Endeavor was also well illustrated at Buffalo. An unusual number of denominational leaders were there, especially the denominational secretaries of young people's work. The denominations are trusting Christian Endeavor. They are looking to it for inspiration and help. They are ready in turn to help Christian Endeavor.

To an extent greater than ever before, the denominational leaders are willing to enter Christian Endeavor councils and guide the movement,—something for which Christian Endeavor leaders have always been eager. The denominational young people's secretaries are by virtue of their offices trustees of the United Society of Christian Endeavor, and at Buffalo the United Society took a long step in advance by forming a Standards and Programme Commission, to be made up of the denominational secretaries of young people's work and the executive officers of the United Society, an expert body whose task it will be henceforth to determine all lines of work in which Christian Endeavor and the denominations are expected to co-operate.

In the meantime, while this new commission is getting to work, the trustees, executive officers, and field-secretaries outlined with care a simple but aggressive and inspiring programme for the next two years. It deals with fundamentals in regard to which all denominations will wish their Endeavorers to advance, and it presents in a strong way some of the most important principles of Christian service. The "goals" of 1920-21 will be more popular and will prove more profitable than any other proposals ever placed before Christian Endeavor.

A formal and definite alliance between the Boy Scouts and Christian Endeavor was proposed at Buffalo, and the first steps toward it were authoritatively taken. The arrangement will be free and flexible, and will be of decided advantage to both bodies.

None of the standard and well-recognized lines of Christian Endeavor activity were forgotten, but all were recognized, stimulated, and developed at Buffalo. The Tenth Legion had a rousing banquet, with a glorious conference on stewardship. The Quiet Hour had constant mention in the meetings, and was illustrated in many devotional addresses, in one of the best open-air sunrise prayer meetings ever held, and in the Conference

Quiet Hours. The citizenship and temperance rally was a rouser. Union work was promoted in many ways, but especially by a new feature, Challenge meetings, in which the various State and local unions "dared" each other to many large undertakings and initiated many contests that will fill the next two years with pep and point.

The Christian Endeavor committees, the prayer meetings and socials, were by no means forgotten, but were discussed with zest which proved that the Endeavorers are anything but weary of time-honored methods. Missions were to the fore, as always in Christian Endeavor meetings, and both home and foreign missions were illustrated forcibly in pageant, in stereopticon pictures, and in noble addresses. An entire session, and one of the best, was devoted to the New Interchurch World Movement. The enrolment of Life-Work Recruits was promoted quietly but effectively. Not one of our great Christian Endeavor aims and ideals was omitted from the comprehensive programme.

If we were asked to name any one outstanding feature of the Buffalo Conference, we should say, the Alumni banquet. The Alumni idea was proved to have such a hold on the interest and affection of the Endeavorers as ensures its permanence, and with it the continued devotion of the old-time members of our society. Thus a strong backing for Christian Endeavor is made certain, a backing of influence as well as of money. The gift of more than $10,000 made by the 800 Alumni at their banquet is an indication of the way in which this Alumni plan will place Christian Endeavor on an adequate financial basis. The moral support of these old-time Endeavorers is equally important and equally prized. In return, their continued association with the ever-youthful Christian Endeavor society will keep them young, and add indefinitely to the joy and power of their Christian life.

The Des Moines, Ia., delegates held a huckleberry pie social in a down-town restaurant after one of the evening sessions. It was a merry party with a special longing for pie. Mr. Graves and Mr. Davis led the specialties. Treasurer Shartle of the United Society was the guest of the evening.

One of the most enthusiastic delegations at the Conference was the splendid group of more than two hundred from the old Keystone State. Pennsylvania surely was full of pep, and with Reichel, their new field-secretary, Christian Endeavor will indeed continue a leading factor in the life of the Pennsylvania State Christian Endeavor union.

1919-1921 LOYALTY CAMPAIGN.

LOYALTY TO CHRIST.
LOYALTY TO THE C. E. PLEDGE.
LOYALTY TO THE CHURCH.
LOYALTY TO C. E. IDEALS.

LOYALTY IN SERVICE:
(a) Pledged Service.
(b) Outspoken Service.
(c) Systematic Service.
(d) United Service.

Our Standard of Loyalty:

"WHATEVER HE would like to have me do."

WHATEVER:
In Living, Doing, Being, Thinking, Speaking, Giving.

WHATEVER:
In Church, Sunday School and Christian Endeavor.

WHATEVER:
In Home, Business, Community and State.

WHATEVER:
In Recreation, Social Standards and Amusements.

Definite Loyalty Goals on Two-Year Basis:

1. A Christian Endeavor Society ideal of 50 per cent net gain in membership.
2. One or more Christian Endeavor Societies in every available church, or if not, the reason why.
3. A definite goal in every State and Union organization for denominational missionary gifts.
4. A sustaining Alumni Association in every Union.
5. Development of Union organization and Christian Endeavor Field Work.

Challenge, "Come on"—Response, "Let's Go."

MONTHLY SERVICE THEMES, 1919-1920.

September—October

ENLARGEMENT AND ENLISTMENT.
New members for the societies.
New societies: Young People's, Intermediate, Junior, etc.
Citizenship Days; October 12, 1919; October 17, 1920.

November

MISSIONS.
Foreign-Mission Study Classes, Tenth Legion, Life-Work Recruits.

December—January

CHRISTIAN ENDEAVOR PRAYER MEETING.
December: Attendance, Consecration Meeting, Music, Private Devotions, Quiet Hour.
January: Testimony, Public Prayer, Personal Evangelism.

February

MISSIONS.
Home-Mission Study Class, Tenth Legion, Life-Work Recruits.
Christian Endeavor Week, February 1-8, 1920.

March—April—May

CHRISTIAN ENDEAVOR PLEDGE AND COMMITTEE WORK.
March: Church loyalty, church attendance, church support.
April: Lookout Committee, Missionary Committee, Flower Committee, Sunday School Committee, Good-Literature Committee, Junior Commitee.
May: Prayer-Meeting Committee, Social Committee, Music Committee, Information Committee, Citizenship Committee, Temperance Committee, Pastor's Aid Committee.

June

CONVENTION MONTH.
Work of Christian Endeavor Unions.

July—August

OUTDOOR AND COMMUNITY SERVICE.
Country Week, Fresh-Air Campaigns, Hospital Work, Floating Work, Outdoor Evangelism, Summer Conferences.

JUNIOR MONTHLY SERVICE THEMES.

The monthly service programmes are to follow, as nearly as possible, those of the Young People's society.

September: Organization of new societies and Junior Expert classes.

October: Increase in membership. Graduation of Experts.

November: Home missions. New subscriptions to *The Junior Christian Endeavor World.*

December: The Child's Quiet Hour.

January: Tenth Legion.

February: Foreign missions.

March: Bible drill work; church attendance.

April: Junior committee of helpers from the Young People's society.

May: Graduation into the Intermediate society.

June: Whatsoever (States' or societies' particular need).

July-August: Outdoor and community service.

After the conference the workers had supper together, and a social half-hour was enjoyed.

Outline for Christian Endeavor Week Programme, 1920
Christian Endeavor Day

Sunday, February 1. Special programme to be issued by the United Society of Christian Endeavor. Topic: "What Does Our Pledge Require?"

Decision Day

Sunday, February 8. Special programme to be issued by the United Society. Topic: "A Worth-While Life."

Note.

It is suggested that societies form a programme of the week's activities, arranging the days of the week to meet local conditions. For instance, Church Loyalty Day should be placed on the day on which the church prayer meeting is held. Fellowship Day may be celebrated on any day of the week. Saturday is suggested for Junior Day. Some unions may want to make Saturday also Intermediate Day. The following are the days suggested, the arrangement of them in the week being left to the unions and societies. Programmes for Fellowship Day, Junior Day, and Intermediate Day will be issued in pamphlet form by the United Society of Christian Endeavor, Boston.

Fellowship Day.

For this day a union rally is suggested. Where there is a Christian Endeavor Alumni Association in the union, a supper and reunion may be held before the rally. Where there is no such association, a Christian Endeavor reunion supper may be held and a

Christian Endeavor Alumni Association organized. In the union rally vital topics of the day should be emphasized, world peace, temperance, prohibition-enforcement, the inter-church movement, and so on.

Church Loyalty Day.

Celebrate this day on the day of the church prayer meeting. Hold a union prayer meeting with the church. The prayer-meeting committee may work out with the pastor a programme for the meeting, giving the Endeavorers a definite part in it. Urge church members who at one time have been Endeavorers to come and tell in the meeting what Christian Endeavor has meant for them. Have the Endeavorers canvass the church members for a full attendance.

Junior Day.

Junior Day may be celebrated on Saturday, February 7. Invite the mothers of the Juniors to this meeting, and in the rally present a model Junior meeting to show the audience how such a meeting is conducted. Have Bible drills; let the Juniors take part in sentence prayers, or by giving talks, object-talks, blackboard-talks, and illustrations. Let them also present a model business meeting, giving their reports. Have Juniors ready to explain the Child's Quiet Hour, Tithing, the Birthday Book, the Training Chart, the Honor Roll, contests, and the work of the committees.

Intermediate Day.

This day may be celebrated on Saturday, February 7, or on any other day of the week. Perhaps a rally on Sunday afternoon might be best. An appropriate week-day celebration would be a banquet for the Intermediates and prospective members, the banquet to be followed by a short, snappy rally for the general public, this rally, in turn, to be followed by a social hour. For a Sunday celebration have a conventionette, the Intermediates having charge of the meeting.

Social Day.

On this day have a social to which should be invited the officers of the church and Sunday school, Sunday-school teachers and former Endeavorers. Messages should be secured from absent members and read in the meeting.

Extension Day.

This day should be the culmination of weeks, perhaps months, of work. The membership of most societies could be doubled. Work for the enlargement of your membership. Do not forget to seek new associate members, affiliated members and honorary members. The number of societies could be greatly increased if societies would plan to organize new societies. Seek out nearby churches that have no Christian Endeavor society; invite their young people to come and see one of your meetings. Go to them, after consulting with their pastor, in a body, and help them to organize a society. Organize Junior and Intermediate societies where there are none in your own church or in other churches.

Note.

Programmes for Sunday, February 1 (Christian Endeavor Day), and for February 8 (Decision Day), will be sold at $2.00 a hundred. Single copy, 5 cents, postpaid. This price is net.

Programmes for Fellowship Day, Junior Day, and Intermediate Day will be sold at the rate of $1.50 a hundred. Single copy, 3 cents, postpaid. This price is net. This folder of programmes is sold at $2.00 a hundred. Single copy, 3 cents, postpaid.

United Society of Christian Endeavor,

Boston, Mass. **Chicago, Ill.**

OFFICERS OF THE UNITED SOCIETY OF CHRISTIAN ENDEAVOR

President.
REV. FRANCIS E. CLARK, D.D., LL.D.

Vice-President.
REV. HOWARD B. GROSE, D.D.

General Secretary.
MR. E. P. GATES.

Treasurer and Publication Manager.
MR. A. J. SHARTLE.

Manager Western Office of Publishing Department.
MR. R. A. WALKER.

Departmental Secretary.
MR. CARROLL M. WRIGHT.

Editorial Secretary.
REV. ROBERT P. ANDERSON.

Associate President and Citizenship Superintendent.
DANIEL A. POLING, LL.D.

Extension Secretary.
REV. IRA LANDRITH, D.D., LL.D.

Chairman All-South Extension Committee.
LIEUT. DUNCAN B. CURRY.

Southwestern Federation Secretary.
MR. W. ROY BREG.

Clerk.
MR. H. N. LATHROP.

Alumni Superintendent.
MR. STANLEY B. VANDERSALL.

Superintendent of Rural and Family Endeavor.
MR. JOHN R. CLEMENTS.

National Intermediate Superintendent.
REV. PAUL C. BROWN.

National Junior Superintendent.
MISS GRACE F. HOOPER.

Army and Navy Superintendent.
REV. S. C. RAMSDEN.

Auditor.
MR. J. J. ARAKELYAN.

Office.
WORLD'S CHRISTIAN ENDEAVOR BUILDING, MOUNT VERNON STREET, BOSTON.

Denominational Trustees.

Rev. Ernest Bourner Allen, D.D.,	Congregational.
Rev. B. W. Arnett, D.D.,	African Methodist Episcopal.
Rev. Thomas Ashburn, D.D.,	Cumberland Presbyterian.
John Willis Baer, LL.D.,	Presbyterian.
Mr. Fred L. Ball,	Congregational.
Rev. W. H. Barraclough, B.A.,	Methodist Church of Canada.
Rev. J. Whitcomb Brougher, D.D.,	Baptist.
Rev. David James Burrell, D.D.,	Reformed Church in America.
Rev. W. W. Bustard, D.D.,	Baptist.
Rev. Julian C. Caldwell, D.D.,	African Methodist Episcopal.
Rev. G. C. Carpenter,	Brethren.
Rev. Ora W. Carrell,	Friends.
Mr. George W. Coleman, M.A.,	Baptist.
Rev. J. C. Cummins,	General Baptist.
Rev. W. J. Darby, D.D.,	Presbyterian
Rev. Frank J. Day, D.D.,	Congregational of Canada.
Rev. H. A. Denton, D.D.,	Disciples of Christ.
Rev. J. Stanley Durkee, D.D.,	Congregational.
Bishop Samuel Fallows, D.D., LL.D.,	Reformed Episcopal.
President Harlan L. Feeman, D.D.,	Methodist Protestant.
Bishop J. S. Flipper,	African Methodist Episcopal.
Rev. William Hiram Foulkes, D.D.,	Presbyterian.
Rev. T. M. Funk,	Church of God.
Rev. J. H. Garrison, LL.D.,	Disciples of Christ.
Rev. Howard B. Grose, D.D.,	Baptist.
Rev. N. B. Grubb, D.D.,	Mennonite.
Mr. William Phillips Hall,	Methodist Episcopal.
Rev. A. W. Halsey, D.D.,	Presbyterian.
Rev. P. A. Heilman, D.D.,	Lutheran.
Professor James Lewis Howe, Ph.D.,	Southern Presbyterian.
Rev. Elijah Humphries, D.D.,	Primitive Methodist.
Rev. Albert W. Jefferson, D.D.,	Free Baptist.
Rev. W. T. Johnson, D.D.,	African Baptist.
Rev. Henry N. Jordan,	Seventh Day Baptist.
Rev. Gilby C. Kelly, D.D.,	Methodist Episcopal Church South.
President Henry Churchill King, D.D., LL.D.,	Congregational.
Bishop Lynwood Westinghouse Kyles, A.M., D.D.,	African Methodist Episcopal Zion.
Rev. Ira Landrith, D.D., LL.D.,	Presbyterian.
Rev. Cleland B. McAfee, D.D.,	Presbyterian.
Rev. J. T. McCrory, D.D.,	United Presbyterian.
Rev. Arthur Meilicke,	Moravian.
Rev. Rufus W. Miller, D.D.,	Reformed Church in United States.
Rev. William Patterson, D.D.,	Presbyterian Church of Canada.
Rev. MacLeod M. Pearce, D.D.,	Reformed Presbyterian.
Rev. W. H. Peck, D.D.,	African Methodist Episcopal.
Rev. Allan B. Philputt, D.D.,	Disciples of Christ.
Rev. W. F. Richardson, D.D.,	Disciples of Christ.
Rev. H. F. Shupe, D.D.,	United Brethren.
Rev. Egbert W. Smith, D.D.,	Southern Presbyterian.
Bishop U. F. Swengel, D.D.,	United Evangelical.
Rev. Floyd W. Tomkins, S.T.D.,	Episcopal.
Rev. Hugh K. Walker, D.D.,	Presbyterian.
Honorable John Wanamaker,	Presbyterian.
President E. A. Watkins, D.D.,	Christian Convention.
Bishop W. M. Weekley, D.D.,	United Brethren.
Rev. Earle Wilfley, D.D.,	Disciples of Christ.
Rev. Herbert L. Willett, D.D.,	Disciples of Christ.
Rev. Samuel H. Woodrow, D.D.,	Congregational.

Special Trustees Representing Denominational Young People's Work.

Rev. Edward S. Bromer, D.D.,	Reformed Church in U. S.
Professor Aaron Brown,	A. M. E. Zion, Varick League.
Rev. O. T. Deever,	United Brethren.
Rev. Gilbert Glass, D.D.,	Presbyterian Church in U. S.
Rev. G. W. Haddaway, D.D.,	Methodist Protestant.
Rev. William Ralph Hall,	Presbyterian Church in U. S. A.
Rev. A. B. Kendall, D.D.,	Christian.
Rev. W. E. Peffley,	United Evangelical.
Rev. Frank M. Sheldon, D.D.,	Congregational.
Mr. John D. Zimmerman,	Disciples of Christ.

Christian Endeavor Field Secretaries.

All the workers in this list are employed by their States and give full-time service.

California,	Mr. Paul C. Brown, 210 Bradbury Building, Los Angeles.
Colorado,	Miss Ida C. Clothier, Manitou.
Illinois,	Mr. F. D. G. Walker, 405 Association Building, Chicago.
Indiana,	....
Iowa,	Miss Mary B. McPherrin, 1514 N. 40th St., Omaha, Neb.
Michigan,	Mr. A. LaVerne Spafford, 1924 Horton Ave., Grand Rapids.
Minnesota,	Miss Margaret M. Bradt, 2383 Long Ave., St. Paul.
Nebraska,	Miss Vida Leamer, Wakefield.
New York,	Mr. Carlton M. Sherwood, 181 Franklin St., Buffalo.
Ohio,	Mr. Arthur E. Whitney, 601 Chamber of Commerce, Columbus.
	Miss Sylvia Tschantz, 601 Chamber of Commerce, Columbus.
Pennsylvania,	Mr. Haines A. Reichel, 1549 State St., Harrisburg.
Southern States,	Mr. Karl Lehmann, 5 Ferger Building, Chattanooga, Tenn.
	Mr. Charles F. Evans, 424 W. High St., Lexington, Ky.
	Rev. L. C. Little, 5 Ferger Building, Chattanooga, Tenn.
Southwestern Federation,	Mr. W. Roy Breg, 1206 Great Southern Life Bldg., Dallas, Texas.
Texas,	Miss Jessie Grider Butts, 1206 Great Southern Life Bldg., Dallas.
West Virginia,	Rev. C. E. Hetzler, 397 Broaddus Ave., Clarksburg.
Wisconsin,	Rev. Edgar T. Farrill, 100 32d St., Milwaukee.

Denominational Christian Endeavor Secretaries.

A. M. E.: Rev. Julian C. Caldwell, D.D., Secretary, Nashville, Tenn.
A. M. E. Zion: Professor Aaron Brown, D.D., Pensacola, Fla.
Christian: Rev. A. B. Kendall, D.D., Ravena, N. Y.
Congregational: ....
Disciples of Christ: Mr. John D. Zimmerman, 703 Jackson St., Topeka, Kan.
Friends: Mr. Clarence E. Pickett, 101 South Eighth St., Richmond, Ind.
Methodist Protestant: Rev. G. W. Haddaway, D.D., 2504 Garrison Ave., Baltimore, Md.; Rev. A. G. Dixon, Young People's Secretary, High Point, N. C.
Presbyterian U. S. A.: Rev. William Ralph Hall, Secretary, Witherspoon Building, Philadelphia, Penn.
Presbyterian U. S. (Southern): Rev. Gilbert Glass, D.D., Superintendent, Richmond, Va.
Reformed in U. S.: Rev. Edward S. Bromer, D.D., Greensburg, Penn.
United Brethren: Rev. O. T. Deever, Secretary, Dayton, O.
United Evangelical: Rev. W. E. Peffley, Evangelical Bldg., Harrisburg, Penn.

The Christian Endeavor World.

Honorary Editor, Rev. Francis E. Clark, LL.D.
Editor, Amos R. Wells, Litt.D.
Associate Editors, Arthur W. Kelly and Rev. Robert P. Anderson.
Contributing Editor, Rev. Daniel A. Poling, LL.D.
Editor Afield, Rev. Ira Landrith, LL.D.
Publisher, William Shaw.
Field Representative, C. C. Hamilton.
The price of *The Christian Endeavor World* is $2.00 a year, in advance, for a domestic subscription; two yearly subscriptions, one of which must be new, for $3.25. Five cents a copy. Canadian subscription, $2.00 net. To any country in the Universal Postal Union, 10s. 3d., or $2.50 a year, postpaid.

THE WORLD'S CHRISTIAN ENDEAVOR UNION

World's Christian Endeavor Building, Mount Vernon Street, Boston, Mass.

President.
Rev. Francis E. Clark, D.D., LL.D.

Vice-President.
Daniel A. Poling, LL.D.

Honorary Secretary.
John Willis Baer, LL.D.

Office Secretary.
William Shaw, LL.D.

Treasurer.

Mr. Stanley B. Vandersall.

Auditor.

Mr. J. J. Arakelyan.

Trustees in America.

Rev. R. P. Anderson.
Mr. J. J. Arakelyan.
Rev. J. L. Barton, D.D.
Mr. John R. Clements.
Pres. J. Stanley Durkee, D.D.
Rev. Howard B. Grose, D.D.
Mr. William Phillips Hall.
John P. Hartman, Esq.
Hon. J. A. McDonald, LL.D.
Mr. Walter R. Mee.
Daniel A. Poling, LL.D.
Mr. A. J. Shartle.
Rev. Charles M. Sheldon, D. D.
Rev. F. G. Coffin, D.D.
Rev. James L. Hill, D.D.
Mr. William N. Jenkins.
Mr. Arthur W. Kelly.
Rev. Ira Landrith, D.D., LL.D.
Mr. H. N. Lathrop.
H. L. McClusky, M.D.
Mr. Charles G. Stewart.
Hon. W. J. Van Patten.
Rev. Von Ogden Vogt.
Prof. Amos R. Wells, LL.D., Litt.D.

Trustees in Foreign Lands.

Australia: Mr. John B. Spencer, Burwood, New South Wales.
Brazil: Rev. Erasmo deC. Braga, Campinas.
Burma: Rev. L. T. Ah Syoo, Moulmein.
Ceylon: Mr. George Nahanielsz, Colombo.
China, Northern: Rev. Paul L. Corbin, Taikuhisien, Shansi.
China, Central: Rev. P. F. Price, Nanking.
France: Mme. M. Lepoids, 123 Ave du Maine, Paris.
Germany: Rev. Friedrich Blecher, Friedrichshagen (bei Berlin).
Great Britain: Mr. James Kelly, 16 Loch Leven Road, Langside, Glasgow, Scotland.
Great Britain: Rev. John Fleming, Sheffield, England.
Great Britain: Rev. F. J. Horsefield, St. Silas Vicarage, Bristol, England.
Great Britain: Professor G. Currie Martin, Valima, 1 Ravenscroft Ave., Golder's Green, N. W. London.
Great Britain: Rev. John Pollock, Belfast, Ireland.
India: Rev. R. A. Hume, D.D., Ahmednegar.
Jamaica, West Indies: Rev. J. E. Randall, 70 Duke St., Kingston.
Japan: Rev. Tasuku Harada, President, Doshisha College, Kyoto.
Mexico: Rev. S. Guy Inman, New York, N. Y.
South Africa: Miss M F. Baldwin, Wellington.
Spain: Rev. William H. Gulick, D.D. (Sarria), Barcelona.

Foreign Vice-Presidents.

Australia: Rev. R. I. Williams.
Brazil: Mario Pinto de Souza Neves, R. Travares Guerra, 77 c x, Ponta do Cajn, Rio de Janeiro.
Bulgaria: Rev. T. Holway, Sofia.
Burma: Rev. H. I. Marshall, Tharrawaddy, Burma.
Chile: Dr. W. E. Browning, D.D., Casilla 2037, Santiago.
China: Rev. G. F. Fitch, American Presbyterian Mission, Shanghai.
Croatia: Mr. Samuel Schumacher, Agram.
Egypt: Miss Anna Y. Thompson, Cairo.
Finland: Miss Aina Johansson, 29 Hogbergsgaten, Helsingfors.
France: Pastor C. E. Greig, 18 Rue de Wattignies, XII, Paris.
Great Britain, England: Rev. W. Bainbridge, 42 Tolver Road, Penzance.
Great Britain, Ireland: Mr. W. M. Clow, J.P., Federal House, Portadown.
Great Britain, Scotland: Mr. James Sneddon, 76 Park Drive, Glasgow.
Great Britain, Wales: Professor W. T. Chance, B.D., Baptist College, Cardiff.
Greece: Rev. J. R. House, D.D., Salonica.
Hungary: Professor A. Szabo, Kalvinter 7, Budapest.
India: Rev. Herbert Anderson, 48 Rippon St., Calcutta.
Italy: Rev. James Gordon Gray, D.D., Via X X Settembre 7, Rome.
Jamaica, West Indies: Rev. W. Marwick, Falmouth.
Japan: Rev. T. Makino, Kyoto.
Mexico: Rev. Leandro Garzo Mora, Montemorelos, Box 6, N. L.
Norway: Rev. H. B. Klaboe, Christiania.
Servia: Miss Mary Matthews, Monastir.
South Africa: Mr. J. R. Farquhar, 30 Warren St., Cape Town.
Spain: Don Carlos Araujo, Jr., Monteleon, 13, 2°, Madrid.
Sweden: Professor Ad. Kolmodin, Upsala University.

American Vice-Presidents.

Each State and Provincial Union is entitled to name one Vice-President to serve for four years or until his successor is appointed. The following is the list to date:

Alabama.	Mr. H. O. Bernard.
California,	Mr. W. N. Jenkins.
Florida.	Mr. Marcus C. Fagg.
Georgia,	Rev. A. R. Moore, Ph.D.
Indiana.	Rev. H. B. Hostetter.
Iowa,	Rev. C. S. Medbury, D.D.
Kansas,	Mr. Paul M. Williams.
Kentucky,	Rev. Homer W. Carpenter.
Maine,	Rev. Harold S. Capron.
Maryland,	Mr. Spencer E. Sisco.
Massachusetts,	Henry L. McClusky, M.D.
Minnesota,	Mr. C. B. Whittier.
Missouri,	Mr. J. Blaine Noggle.
New Hampshire,	Mr. Edward A. Dame.
New Jersey,	Mr. W. Egbert Thomas.
New York,	Rev. F. G. Coffin, D.D.
North Carolina,	Mr. H. Galt Braxton.
North Dakota,	Mr. George L. Hempstead.
Ohio.	Rev. Charles R. Sine.
Oklahoma,	Miss Athie E. Sale.
Oregon.	Mr. G. Evert Baker.
Pennsylvania.	Rev. R. G. Bannon, D.D.
Rhode Island,	Miss Muriel McFee.
South Carolina,	Mr. Allan Nicholson.
Texas,	Rev. Patrick Henry.
Virginia,	Mr. G. E. Miley.
West Virginia,	Mr. Louis E. Black.
Wisconsin,	Rev. E. T. Farrill.

CHRISTIAN ENDEAVOR DIRECTORY

STATE AND PROVINCIAL UNIONS

International C. E. Field Workers' Union.

Mr. E. P. Gates, Pres., 41 Mt. Vernon St., Boston, Mass.
Mr. Haines A. Reichel, Vice-Pres., 1549 State St., Harrisburg, Pa.
Miss Sylvia Tschantz, Sec., 601 Chamber of Commerce, Columbus, Ohio.
Mr. F. D. G. Walker, Treas., 405 Association Building, Chicago, Ill.

All-South Extension Committee of the United Society of C. E.

Headquarters, 5 Ferger Bldg., Chattanooga, Tenn.
Lieut. Duncan B. Curry, Chairman, Jacksonville, Fla.
Mr. Wyatt A. Taylor, Sec., Columbia, S. C.
Mr. John M. Gore, Vice-Chairman, Box 195, Knoxville, Tenn.
Rev. L. E. Brubaker, Vice-Chairman, Ensley, Ala.
Mrs. Wyatt A. Taylor, Treasurer, 1316 Gervais St., Columbia, S. C.
Southern States Secretary,
Mr. Karl Lehmann, 5 Ferger Bldg., Chattanooga, Tenn.
Field Secretaries,
Mr. Charles F. Evans, 424 W. High St., Lexington, Ky.
Rev. L. C. Little, 5 Ferger Bldg., Chattanooga, Tenn.

Southwestern Christian Endeavor Federation.

Rev. Francis E. Clark, D.D., LL.D., Honorary Chairman, 41 Mt. Vernon St., Boston, Mass.
Mr. Ralph S. Young, Chairman, 401 Mining Exchange Bldg., Colorado Springs, Col.
Miss Emma McClaran, Vice-Chairman, Tulsa, Okla.
Miss Lillian E. Dyer, Vice-Chairman, 803 Fourth St., Alva, Okla.
Rev. J. B. Gonzales, Treas., 4532 Live Oak St., Dallas, Tex.
Mr. W. Roy Breg, Southwestern Secretary, 1206 Great Southern Life Bldg., Dallas, Tex.

UNITED STATES.

Alabama.

Mr. A. Clinton Decker, Pres., P. O. Box 790, Birmingham.
Mr. E. S. Clark, Sec., 325 S. 6oth St., Birmingham.
Miss Bessie Wolf, Junior Supt., 1606 13th Ave., S., Birmingham.
Mr. H. L. Henley, Intermediate Supt., Chamber of Commerce, Albany.
Miss Lillie Gafford, Supt. Introduction Dept., Montgomery.

Alabama (Colored).

......................, Pres., ...
Mr. L. C. Farley, Gen. Sec., Beloit.
Mr. Edward D. Nelson, Rec. Sec., Miller's Ferry.
Mrs. M. A. Dillard, Junior Supt., 1430 Ann St., Selma.

Alaska.

*Rev. Edward Marsden, Supt.,
Saxman, Ketchikan, P. O.

Arizona.

Mr. Howard M. Patterson, Supt.,
P. O. Box 626, Tucson.
Miss Verna Harvey, Junior Supt.,
723 N. Pierce St., Phœnix.

Arkansas.

Miss Mary F. Bullock, Pres.,
709 W. Main St., Russellville.
Mrs. Elwin Shipley, Sec.,
Fort Smith.
Miss Pearl Holloway, Junior Supt.
Farmington.
Mrs. Hattie Carmichael,
Intermediate Supt.,
Fort Smith.

California.

Mr. Ben J. Small, Pres.,
1735 19th Ave., Oakland.
Mr. B. Irvin Valentine, Gen. Sec.,
555 9th St., San Bernardino.
Miss Lulu M. Minter,
Cor. and Stat. Sec.,
322 W. 3d St., Santa Ana.
Miss Louise Sommers,
Junior and Intermediate Stat. Sec.,
834 3d St., Santa Rosa.
Mr. Paul C. Brown, Field Sec.,
210 Bradbury Bldg., Los Angeles.
Mrs. Faith Voorhies Power,
Junior Supt.,
Livermore.
Mr. Ray H. Horton, Intermediate Supt.,
3931 Walton Ave., Los Angeles.
Miss Elsa Koch,
Supt. Introduction Dept.,
3227 W. St., Sacramento.

Colorado.

Mr. Ralph S. Young, Pres.,
401 Mining Exchange Bldg.,
Colorado Springs.
Miss Julia Lindstrom, Gen. Sec.,
3320 Zuni St., Denver.
Miss Ida C. Clothier, Field Sec.,
Manitou.
Miss Grace M. Torbet, Junior Supt.,
727 W. 13th St., Pueblo.
Mr. Julius Tilquist, Intermediate Supt.,
82 W. Byers St., Denver

Connecticut.

Rev. Morris E. Alling, Pres.,
Box 164, Rocky Hill.
Miss Cora K. Edell, Sec.,
26 Maplewood Ave., Hartford.
Mrs. Leon Everett Taylor,
Junior and Intermediate Supt.,
....

Delaware.

Dr. Evans Roberts, Pres., Milford.
Miss Dorothy H. Krauss, Sec.
Henry Clay.
Miss Sadie McKinney,
Junior and Intermediate Supt.,
16th and Riverview Ave.,
Wilmington.

District of Columbia.

Mr. D. F. Michael, Pres.,
1325 G St., N. W., Washington.
Miss Mary E. Close, Sec.,
1325 G St., N. W., Washington.
Miss Kathryn Thompson, Junior Supt.,
1325 G St., N. W., Washington.
Miss Minnie Nicholson,
Intermediate Supt.,
1325 G St., N. W., Washington.

Florida.

Lieut. Duncan B. Curry, Pres.,
405 Atlantic National Bank Bldg.,
Jacksonville.
Miss Katherine Bishop, Sec.,
8 Vernon Terrace, Jacksonville.
Miss Elizabeth Weimar, Junior Supt.,
Fernandina.
Miss Lucy Harris, Intermediate Supt.,
Bradentown.
Miss Constance Harris,
Supt. Introduction Dept.,
St. Cloud.

Georgia.

Mr. George R. Rusk, Pres., Decatur.
Miss Nora Saye, Sec.,
P. O. Box 534, Athens.
Mrs. J. H. Moye, Junior Supt.,
Atlanta.
Mr. Russell M. Bythewood,
Intermediate Supt., St. Marys.

Hawaiian Islands.

Rev. Akaiko Akana, Pres.,
P. O. Box 489, Honolulu.
Mr. P. D. Kellett, Jr., Sec.,
211 McCandless Bldg., Honolulu.

Idaho.

Mrs. J. Q. Hook, Pres.,
P. O. Box 271, Boise.
Miss Gertrude Shoemaker, Sec.,
719 9th Ave. S., Nampa.
Mr. Marion Smyth, Junior Supt.,
P. O. Box 694, Nampa.
Mrs. A. E. Johnson,
Intermediate Supt., Rupert.
Miss Winifred Bossen,
Supt. Introduction Dept., Fruitland.

Illinois.

Rev. I. W. Bingaman, Pres.,
413 S. 12th St., Quincy.
Miss G. Geneva Doran, Sec.,
6435 Kenwood Ave., Chicago.
Mr. F. D. G. Walker, Field Sec.,
405 Association Bldg., Chicago.
Mrs. William V. Martin, Junior Supt.,
42 West St., Freeport.
Miss Laura Belle Stephens,
Intermediate Supt.,
405 Association Bldg., Chicago.
Miss Janet Weir,
Supt. Introduction Dept.,
4310 Van Buren St., Chicago.

Indiana.

Mr. O. F. Gilliom, Pres., Berne.
Miss Mary Roy Thomson, Sec.,
4140 Broadway, Indianapolis.
Miss Ruth Day, Junior Supt., Marion.
Miss Bessie Barrows, Intermediate Supt.,
1112 Chandler Ave., Evansville.
Rev. Harold Martin,
Supt. Introduction Dept., Winimac.

Iowa.

Rev. C. H. Stauffacher, Pres.,
408 A Ave., West, Cedar Rapids.

*Appointed by U. S. C. E.

Miss Mary B. McPherrin,
General and Field Sec.,
1514 N. 40th St., Omaha, Neb.
Mrs. S. N. Ramiage, Junior Supt.,
7 N. 5th St., Marshalltown.
Miss Jessie L. Morrow,
Intermediate Supt.,
1406 Riverside Ave., Sioux City.

Kansas.

Mr. R. E. Hangen, Pres., Wellington.
Miss Madilene Carter, Cor. Sec.,
Lawrence.
Rev. J. W. Shell, General Sec.,
804 Lafayette, Kansas City.
Mr. Walter L. Clark, Acting Field Sec.,
Emporia.
Mrs. J. H. Hindman, Junior Supt.,
Humboldt.
Miss Anna T. Willis, Intermediate Supt.,
808 W. 9th St., Emporia.
Miss Jeanne Severance,
Supt. Introduction Dept.,
Garden City.

Kentucky.

Rev. W. H. Hooper, Pres.,
3033 Virginia Ave., Louisville.
Miss Mary L. Wilson, Sec.,
34 Limestone St. ,Maysville.
Mr. Charles F. Evans, Field Sec.,
424 W. High St., Lexington.
Miss Carrie F. Adams, Junior Supt.,
2232 Madison Ave., Louisville.
Miss Anna Belle Stewart,
Intermediate Supt., Mt. Vernon.
Mrs. Charles F. Evans,
Supt. Introduction Dept.,
424 W. High St., Lexington.

Louisiana.

Mr. J. L. Webster, Pres.,
Stonewall St., Shreveport.
Miss Mary T. Pirie, Sec.,
222 St. Napoleon St., Baton Rouge.
Rev. L. C. Little, Field Sec., Winnfield.
Miss Louise Kemp, Junior Supt.,
920 Olive St., Shreveport.
Mrs. T. M. Millican,
Intermediate Supt., Bastrop.

Maine.

Rev. Harold S. Capron, Pres.,
96 4th St., Bangor.
Miss Lena Kerr, Sec., Oakland.
Miss Sarah Morgan,
Junior and Intermediate Supt.,
Oak St., Lewiston.
Miss Ethel M. Burgess,
Supt. Introduction Dept.,
126A Danforth St., Portland.

Maryland.

Mr. William I. Deter, Pres.,
Myersville.
Mr. Spencer E. Sisco, Gen. Sec.,
1605 Continental Bldg., Baltimore.
Mr. George R. Smith, Junior Supt.,
1836 E. North Ave., Baltimore.
Miss Dora M. Lohmeyer,
Intermediate Supt.,
922 N. Gilmor St., Baltimore.
Miss Ethel L. Fowler,
Supt. Introduction Dept.,
725 Dolphin St., Baltimore.

Massachusetts.

Henry L. McClusky, M.D., Pres.,
7 Hawthorne St., Worcester.
Mr. Karl F. Wilson, Gen. Sec.,
167 Highland St., Worcester.
Mr. Asa H. Burgess, Rec. Sec.,
29 Samoset St., Plymouth.
Miss Alice L. Mason, Junior Supt.,
47 Riddell St., Greenfield.
Mr. Ronald R. Johnson,
Intermediate Supt.,
33 Florence St., Worcester.
Miss Grace E. Davis,
Supt. Introduction Dept.,
North Wilmington.

Michigan.

Mr. Walter C. Radeke, Pres.,
Saginaw, W. S.
Mr. A. LaVerne Spafford,
Rec. and Field Sec.,
1924 Horton Ave., Grand Rapids.
Miss Ethel A. Brown, Junior Supt.,
112 Woodworth St., Bad Axe.
Miss C. Edna Lewis, Intermediate Supt.,
241 Massachusetts Ave., Detroit.
Miss Ruth Symons,
Supt. Introduction Dept.,
716 S. Weadock Ave., Saginaw.

Minnesota.

Mr. C. B. Whittier, Pres., Farmington.
Miss Anna L. Hartin, Sec., Hastings.
Miss Margaret M. Bradt, Field Sec.,
2383 Long Ave., St. Paul.
Miss Edna M. Fellows, Junior Supt.,
Worthington.
Miss Ethel Boxell, Intermediate Supt.,
1397 Hewitt Ave., St. Paul.
Miss Bessie O'Brien,
Supt. Introduction Dept.,
416 Broadway, N. E., Minneapolis.

Mississippi.

Rev. John A. Wood, Pres., Canton.
Miss Maude Marks, Rec. Sec., Jackson.
Miss Myra Hume, Cor. Sec., University.
Miss Antoinette Clark, Junior Supt.,
Tupelo.
Miss Alma Sims, Intermediate Supt.
Oxford.

Missouri.

Mr. Karl E. Lubke, Pres.,
Suite 813, Fullerton Bldg., St. Louis.
Miss Charlotte Schilling, Sec.,
902 Federal Reserve Bank Bldg.,
St. Louis.
Miss Nellie Parsons, Stat. Sec.,
514 Normal Ave., Warrensburg.
Miss Dora Clemens, Junior Supt.,
5050 Cates Ave., St. Louis.
Miss Gertrude Brooks,
Intermediate Supt.,
867 Newport Ave., Webster Groves.
Mrs. John Medcalf,
318 E. 1st St., Monroe City.

Montana and Northern Wyoming.

Mr. A. Paul Thompson, Pres.,
11 23d St., N., Great Falls.
Miss Mabel Gaylord, Sec.,
P. O. Box 1043, Great Falls.
Mrs. Floyd Ghormley, Junior Supt.,
Kalma.
Mrs. C. H. Lorimer, Intermediate Supt.,
P. O. Box 501, Deer Lodge.

Nebraska.

Rev. C. C. Dobbs, Pres., Aurora.
Miss Addie Wagey, Gen. Sec.,
Cambridge.

Miss Vida Leamer, Field Sec., Wakefield.
Miss Fern Betts, Junior Supt., Franklin.
Miss Alta Harrison, Intermediate Supt., 1319 15th St., Aurora.

Nevada.

*Mr. J. M. Ellis, Supt., Tonopah.

New Hampshire.

Mr. Edward A. Dame, Pres., 6 School St., Concord.
Miss Elizabeth A. Dodge, Sec., Andover.
Mrs. William C. Landis, Junior Supt., Lakeport.
Mrs. Leon B. Proctor, Intermediate Supt., Hillsboro.
Miss Hazel P. Adams, Supt., Introduction Dept., 67 Riddle St., Nashua.

New Jersey.

Mr. John T. Sproull, Pres., Arlington.
Miss Nellie K. Foote, Sec., 906 Grove St., Elizabeth.
Miss Mae Hice, Junior Supt., 13 2d Ave., E. Roselle.
Miss Nina Rowland, Intermediate Supt., Clinton.
Miss Helen O. Parsons, Supt. Introduction Dept., P. O. Box 1333, Belmar.

New Mexico.

Rev. E. E. Mathes, Pres., Artesia.
Miss Edna Rogers, Cor. Sec., 214 E. Gold Ave., Albuquerque.
Miss Rhoda Swayne, Rec. Sec., 1406 N. 5th St., Albuquerque.
Rev. C. C. Cordova, Spanish Sec., East Las Vegas.
Miss Bertha Bartlett, Junior and Intermediate Supt., 505 N. Richardson Ave., Roswell.
Miss Alberta Harper, Supt. Introduction Dept., Alamagordo.

New York.

Mr. Fred C. Collins, Pres., 205 N. Salina St., Syracuse.
Mr. W. Nelson Ells, Sec., 135 W. 52d St., New York City.
Mr. Carlton M. Sherwood, Field Sec., 181 Franklin St., Buffalo.
Miss Agnes E. Baker, Junior Supt., 104 Normal Ave., Buffalo.
Rev. George H. Scofield, Intermediate Supt., Highland.
Mr. F. G. Volz, Supt. Introduction Dept., 1560 E. 14th St., Brooklyn.

North Carolina.

Rev. J. Vincent Knight, Pres., Greensboro.
Miss Eunice Long, Sec., Greensboro.
Mrs. W. H. Howell, Junior Supt., Wilmington.
Mr. G. M. Beaty, Intermediate Supt., Charlotte.

North Dakota.

Mr .George L. Hempstead, Pres., Jamestown.
Mr. W. E. Dresser, Sec., New Rockford.
Miss Martha E. Kruempel, Associate Sec., New Rockford.
Miss Ruby Schumann, Junior Supt., Bismarck.
Mr. Fred Young, Intermediate Supt., Courtenay.

Ohio.

Rev. Charles R. Sine, Pres., 719 Campbell Ave., Hamilton.
Mr. Arthur E. Whitney, Gen. Sec., 601 Chamber of Commerce, Columbus.
Miss Hazel Harper, Rec. Sec., Moody Bible Institute, Chicago, Ill.
Miss Ethel Snyder, Intermediate Supt., 601 Chamber of Commerce, Columbus.
Miss Sylvia Tschantz, Junior Field Sec., 601 Chamber of Commerce, Columbus.

Oklahoma.

Miss Emma McClaran, Pres., Tulsa.
Miss M. Stella Hinds, Sec., 215 E. 8th St., Oklahoma City.
Miss E. Bernice Carter, Junior Supt., East Enid.
Miss Avis Shirley, Intermediate Supt., 128 East N. W., Miami.
Miss Tephia Folsom, Supt. Introduction Dept., Atoka.

Oregon.

Miss Faye Steinmetz, Pres., 1520 E. Taylor St., Portland.
Miss Elaine Cooper, Sec., 830 E. Morrison St., Portland.
Miss Faye Moorehead, Junior Supt., The Dalles.
Miss Effie Torgerson, Intermediate Supt., Whitney Apartments, Portland.
Miss Ruth Worden, Supt. Introduction Dept., 1085 Belmont St., Portland.

Pennsylvania.

Mr. H. K. Bragdon, Pres., 1203 Union Bank Bldg., Pittsburgh.
Mr. Haines A. Reichel, Gen. Sec., 1549 State St., Harrisburg.
Miss Sadie H. Wiggins, Junior Supt., 6619 Virgina Ave., Ben Avon, Pittsburgh.
Miss Margaret A. Spencer, Intermediate Supt., 38 Exeter St., Pittsburgh.
Miss Clara R. Matthews, Supt. Introduction Dept., Paoli.

Rhode Island.

Miss Muriel McFee, Pres., 550 S. Main St., Woonsocket.
Mr. Frank O. Bishop, Cor. Sec., City Hall, Providence.
Miss Florence L. Brown, Rec. Sec., 404 Orms St., Providence
Miss Grace E. Cole, Junior Supt., Warren.
Miss Harriet Potter, Intermediate Supt., 190 Broadway, Pawtucket.

South Carolina.

Mr. Wyatt A. Taylor, Pres., 1316 Gervais St., Columbia.
Miss Claudia Fraser, Sec., Sumter.
Miss Ada Saunders, Junior Supt., York.
Miss Hannah Plowden, Intermediate Supt., Kingstree.

*Appointed by U. S. C. E.

South Dakota.

...................., Pres.,
Miss Etha Burnham, Sec., Aberdeen.
Mrs. A. H. Seymour, Junior Supt., Aberdeen.
Mrs. W. J. Page, Intermediate Supt., Bradley.

Tennessee.

Mr. Warren Newsum, Pres., 299 S. Dunlap St., Memphis.
Miss Margaret McKenzie, Sec., Nashville.
Mrs. William B. Thorne, Junior Supt., McKenzie.
Miss Olive Matthews, Intermediate Supt., 1024 Mississippi Ave., Memphis.
Miss Sarah Postlethwaite, Supt. Introduction Dept., 101 Island Ave., Chattanooga.

Texas.

Mr. Marsh Callaway, Pres., 3612 Avie St., Houston.
Mr. W. Roy Breg, Gen. Sec., 1206 Great Southern Life Bldg., Dallas.
Miss Jessie Grider Butts, Field Sec., 1206 Great Southern Life Bldg., Dallas.
Mr. D. B. Trice, Rec. Sec., Timpson.
Miss Helen Schram, Office Sec., 1206 Great Southern Life Bldg., Dallas.
Miss Eda Koch, Junior Supt., 1102 S. Alamo St., San Antonio.
Mr. Ernest M. Ligon, Intermediate Supt., T. C. U., Fort Worth.
Miss Jennie M. Perrin, Supt. Introduction Dept., R. F. D. 3, San Antonio.

Utah.

Miss Mina McArthur, Pres., 135 West 3d South St., Salt Lake City.
Miss Erma Snedaker, Sec., 703 St. 11th East, Salt Lake City.
Miss Edna Flandro, Junior Supt., 1541 S. State St., Salt Lake City.
Miss Esther Matson, Intermediate Supt., 1476 S. 4th East St., Salt Lake City.
Miss Carol Cain, Supt. Introduction Dept., 1430 S. Main St., Salt Lake City.

Vermont.

Rev. C. A. Adams, Pres., Danville.
Mrs. Helen Emerson, Sec., Jeffersonville.
Mrs. H. A. Freeman, Junior Supt., Woodstock.

Virginia.

Mr. G. Errett Miley, Pres., Route 6, Box 140, Richmond.
Miss Bessie Ireson, Cor. Sec., 522 Lee St., Bristol.
Mr. W. B. Kerr, Rec. Sec., 1224 Hamilton Terrace, Roanoke.
Miss Annie Wintermeyer, Junior Supt., 801 9th St., S. E., Roanoke.
Miss Edna Brown, Intermediate Supt., Marion.

Washington.

Miss Louella S. Dyer, Pres., 628 Lumber Exchange, Seattle
Mr. Ray W. Clough, Sec., 4145 Arcade Bldg., Seattle.
Mrs. F. C. Allen, Junior Supt., 7433 48th St., S., Seattle.
Mr. William G. Smith, Intermediate Supt., Aberdeen.
Miss Ada Myers, Supt. Introduction Dept., 617 Shannon Ave., Spokane.

West Virginia.

Rev. A. J. C. Bond, Pres., Salem.
Miss Emma Osgood, Cor. Sec., 633 6th St., Huntginton.
Miss Rosa A. Zill, Rec. Sec., Martinsburg.
Rev. C. E. Hetzler, Field Sec., 397 Broaddus Ave., Clarksburg.
Miss Ethel Roush, Jr., Supt., 1408 Spring St., Parkersburg.
Miss Mary Jackson, Intermediate Supt., Jane Lew.

Wisconsin.

Mr. Charles E. Houtkamp, Pres., 377 4th Ave., Wauwatosa.
Miss Allie Kastein, Gen. Sec., 517 E. Jefferson St., Waupun.
Rev. Edgar T. Farrill, Field Sec., 100 32d St., Milwaukee.
Miss Mary B. Stevens, Junior Supt. Durand.
Mr. Erwin Schuttenhelm, Intermediate Supt., 462 17th Ave., Milwaukee.
Miss Allie Kastein, Supt. Introduction Dept., 517 East Jefferson St., Waupun.

Wyoming.

*Miss Mabelle Carmody, Supt., 411 W. 22d St., Cheyenne.

CANADA.

Alberta.

*Mr. James S. Rankin, Supt., 311 Herald Bldg., Calgary.

British Columbia.

Mr. H. H. Hotson, Acting Pres., Dunlevy Ave., Vancouver.
Mr. W. H. Sime, Acting Cor. Sec., 1556 8th Ave., E., Vancouver.
Miss C. McKenzie, Junior Supt., 2314 St. George St., Vancouver.
Mr. W. J. Hogg, Intermediate Supt., 2073 3d Ave., W., Vancouver.

Manitoba.

Mr. Stewart M. Battram, Pres., 174 Arlington St, Winnipeg.
Miss U. S. Jones, Sec., 1103 McArthur Bldg., Winnipeg.
Miss Grace Williams, Junior Supt., 125 Harriett St., Winnipeg.
Miss Grace Salter, Intermediate Supt., 165 Cathedral Ave., Winnipeg.

---

*Appointed by U. S. C. E.

Nova Scotia.

Rev. George A. Christie, Pres., Lansdowne Station.

Mr. John S. Smith, Supt., Truro.

Ontario.

Mr. A. Russell Hewetson, Pres., Brampton.

Mr. A. M. Simpson, Sec. (Eastern), Ottawa.

Miss M. Edna Breithaupt, Junior Supt., 108 Queen St., N. Kitchener.

Miss Elizabeth Hall, Intermediate Supt., Montreal.

Prince Edward Island.

*Mr. Lloyd Shaw, Supt., New Perth.

Quebec.

Rev. David Gardner, Pres., 712 Adam St., Maisonneuve, Montreal.

Miss E. Hall, Sec. and Supt Introduction Dept., 2546 Hutchison St., Outremont, Montreal.

*Appointed by U. S. C. E.

INDEX.

Made in United States
Troutdale, OR
11/03/2026

56845120R00135